Printed by IngramSpark in the United States of America.

First printing, 2024.

Eden's Refuge

PO Box 427.

Prescott, On, K0E 1T0

www.EdensRefuge.ca

Dedication:

This book is dedicated to every designer, permaculture, or otherwise who has tried to uncover the true values and goals of a client. May this be a tool for you and for your clients to help you create the most appropriate design for their context.

Table of Contents

Preface

This book began specifically for my Permaculture design clients. However, through the writing process, it evolved into something suitable for anyone getting ready to design any sustainable project.

The context for all of the examples in this book are focused on getting ready for a permaculture design, but you may interpret it within your own context, be it building your property or building a business.

Working with a permaculture designer can be a big commitment. You may find a designer who is fresh out of design school and willing to work for less than $50 per hour as they gain experience and confidence. However, in general, your designer will charge substantially more. Rates vary, however, in general, $150/hr. is a low rate for a skilled designer, $175-$250/hr. is more common, and I have talked to consultants who are worth every penny of the $1000+ per hour that they charge.

Professional designers are few and far between. This can make it difficult to get more than one quote or to find the right one for. The new designer at $50/hr may be a waste of time and resources and may set you back years on your property but they may also be a well-learned individual who makes excellent observations. A single hour of walking around your property with them may completely change the way you see your property and could set you up for success.

On the other hand, someone charging $250/hr may have a long list of certifications and a massive portfolio of successful projects or they could be someone who took a Permaculture Design Course but doesn't know how to make good observations or create a design that will function properly. They may just know that permaculture design is specialized work and that they can charge a premium for this work.

One of the biggest shortfalls Permaculturalists face is to get hooked on an idea and want to incorporate it into every design. Common examples are swales, herb spirals, food forests, cob ovens, and the heavy use of woodchips to solve every problem.

No single element is suitable for every project. The cob ovens are an example of an element that isn't always suitable because they don't fare well in all climates, especially without protection from the elements.

Swales are an example of elements that are excellent in the right place but that can be devastating in the wrong context. I love swales when used properly. Putting a swale in the right place can completely transform a landscape into a vibrant ecosystem. Putting a swale in the wrong place could result in a horrifying transformation as an entire hillside becomes a mudslide.

If you are building a business there may be elements that you want to include in your business even if they don't make sense from an economical or process perspective.

Whether it's you, or your designer who may get stuck on an idea, it is important to take the time to understand why every element is in a design and also what the potential problems could be.

A good, experienced designer should be able to avoid the pitfalls of obsessing over specific design elements.

Figure 1: JD Van Allen relocating a feral honeybee swarm from a tree.

About the Author

I am a Permaculture Designer from Ontario, Canada who started studying Permaculture in 2014 while living in Australia. I grew up in the garden and in the kitchen. The garden was central to my life as I was more likely to find a meal in the garden than I was to venture inside the house in search of food.

My passion for growing started when I took a trip to Australia. While working on a 40,000-acre cattle station, I came across an article about the $300 home, A super-adobe, or earthbag home. This sent me down the path of sustainable building practices, looking at earthbags, rammed earth, straw bale, buried, passive solar, and Earthship houses.

In the end, I settled on the Earthship, or something like it because of the efficiency (this coming primarily from an economic viewpoint at the time) and because the idea of having a greenhouse as my front hall would allow me to feel a little more alive during our dark winters.

The Earthship led me to learn about composting and livestock. One word kept popping up and the meaning behind it made sense to me. The word was "Permaculture", referring to "permanent culture".

It was about creating systems that would meet our needs by working with nature. This was in stark contrast to the dusty, abused landscape on which I was working.

The land I was on had paddocks, thousands of acres in size, that couldn't be used. Most areas either had water or feed, but only one could support any cattle. There was a great need for permaculture practices in this space.

That land could be restored to a beautiful ecosystem... but I didn't know enough at the time.

I spent the next 10 years studying and practicing permaculture. I read Bill Mollison's Permaculture Designer's Manual cover-to-cover then turned back to page one and started over. I watched countless YouTube videos, and started applying permaculture in my own life.

Stumbling across a video of Stefan Sobkowiak's farm by Possible Media (Possible Media, 2013) showed me what was possible in my own climate. That's when my whole future shifted. I'm not sure how many times I watched that video in the early years. I experienced awe, wonder, and a desire to reproduce what I saw in that 9 minute video.

In 2017, I bought a house and started propagating plants while also increasing the efficiency of the house to make it more comfortable, more ecologically friendly, and more affordable to heat. The backyard was dug up, along with a large collection of buried garbage, and it was replaced with gardens. Eventually, I put up a greenhouse and started a nursery.

Along this journey I studied a great many topics, some came with certifications while others simply provided knowledge and experience. In every case, I went into a deep dive with a combination of internet research, courses, classes, master classes, and experiments or case studies.

Some of the topics include composting, humanure, rainwater harvesting, passive solar greenhouse design, greenhouse management, botany, beekeeping, native pollinators and their habitats, orchard design, grafting, pruning and training trees, orchard management, creating ecosystem ponds, plant propagation, nursery management, bee rescue/removal, creating natural dyes, fermentation, dehydration, solar dehydration, earthworks, landscape design, construction, renovations, renewable energies, timber framing, forestry, and psychology.

As you can tell, I'm a real nerd when it comes to permaculture and all things under that broad umbrella. A significant portion of my winters are spent studying, this is likely to continue for the rest of my life as this is one of the benefits of living in a cold climate. As growers, we are forced to slow down through the winter, this is a time for reflection, planning, and learning.

This book is my contribution to all permaculture designers and to all people who are working on their own projects from designing a homestead to clarifying the direction of a large company that desires to be more ethical and sustainable.

The current stage began in 2022 when I started Eden's Refuge Permaculture Hub and Nursery. This is my "Zone zero" (you will learn about zones in Chapter 3) for all my businesses from permaculture design to bee rescue and timber framing.

Eden's Refuge, as a design business and as a farm, seeks to meet the longings for the things that the Garden of Eden provided: food, security, and community.

Now I am also an author who never intended to be an author. I hope that this book serves as a launching point for many of you as you begin your journey in a new direction.

Chapter One: What is Permaculture?

Permaculture is an often misunderstood term. It is used to describe gardening practices and is associated with a list of specific elements like swales, cob ovens, and herb spirals. Permaculture is much more than that.

Permaculture is a design science that encompasses all areas of life from house designs to politics, and yes, even garden design. It was co-originated by Bill Mollison and David Holmgren with much of the knowledge coming from a combination of native groups from around the world, and through careful observation.

I own a few large textbooks that describe permaculture, the design principles, and techniques. To describe it in depth is far beyond the capacity and intent of this book but I want to give you some basics.

An important note is that Permaculture can be practiced at any scale. It can be practiced on a 40,000-acre ranch and it can be practiced in a tiny condo on the 37th floor in Toronto. It's not all about gardens but about living intentionally within the core ethics that will be explained on the next page. It's about observation, it's about taking care of ourselves, others, and the Earth.

The Prime Directive:

Bill Mollison (Co-creator of Permaculture) opens his text book "Permaculture A Designer's Manual" by discussing that the book is about design but that it is also about values and ethics. These are the first things that Bill wants to talk about yet they get left out of the discussion far too often.

I will take this quote straight from the designer's manual:

> The Prime Directive of Permaculture.
> The only ethical decision is to take responsibility for our own existence and that of our children.
> **Make it now.** (Mollison, Permaculture A Designer's Manual, 1988)

This is a call to action. The call to action is followed by the ethics of permaculture which are to guide the action.

Ethics

Three ethics guide permaculture design, these are the three ethics that I use to guide my business and my own life. If a decision doesn't follow these ethics is it not and cannot be permaculture.

I will start by giving the textbook definition, followed by my definition.

Earth Care:

"Provision for all life systems to continue and multiply." (Mollison, Permaculture A Designer's Manual, 1988)

My wording for this ethic in my own values statement is:

> "The Earth is our home and I believe that its
> health should factor into every decision we

*make. Having a positive impact is the goal, and
having a neutral impact is the minimum we should aim for"*

This ethic means ensuring that our decisions aren't made just for the advancement of humans in the short term.

As a human race, we have the power to destroy the Earth with the click of a few buttons (nuclear weapons) or by going about our days ignorant of the costs (pollution and excessive consumption). We also have the power to completely transform the future of the Earth in positive ways.

By taking the time to consider the needs of the planet and by observing the systems and patterns that the world uses to repair itself we can support those systems through small actions that result in massive changes.

People Care:
"Provision for people to access those resources necessary for their existence." (Mollison, Permaculture A Designer's Manual, 1988)

My wording for this in my own values statement is:

> *"We must create spaces that
> Meet the physical, emotional, and
> spiritual needs of people so we
> all can thrive in good health"*

No decision that we make should cause harm to other people, nor should it take away from their ability to survive.

<u>Fair Share:</u>

"By governing our own needs, we can set resources aside to further the above principles" (Mollison, Permaculture A Designer's Manual, 1988)

My wording for this in my own values statement is:

> *"We live in a world with abundant*
> *resources and abundant need.*
> *Once we harvest all that we need*
> *we can share the surplus to create*
> *a better life for all. Taking a fair share*
> *and fairly sharing what remains."*

Sometimes this means sharing with people, but sometimes it means something as simple as composting your dinner scraps and sharing those nutrients with the Earth instead of sending them to landfills where they cannot decompose properly and do harm to the Earth.

All of our actions should consider all three of these ethics. The designer also shapes their work through the process of observation.

Chapter Two: Observe and Interact

It is said that a Permaculture Designer's most valuable tool is their hammock. This isn't serious but it also isn't really a joke; it has some merit. Fortunately, I have spent over 2000 nights in a hammock so I'm well equipped for design work!

By taking the time to think about a property, and to think about a design, we can avoid the pitfalls of too much haste. The consequences of haste can range from a result that isn't quite what we want to something that must be removed entirely so the project can start again.

How do we make sure we avoid these pitfalls? In addition to the work on values that we will be doing, a lot comes down to observation.

Observation is a powerful tool for learning. Babies spend much of their time observing and you can see their brain working as they learn that whenever they make a big goofy smile or cry loudly enough, they will receive love and attention. Kids observe how their parents act to learn what is right and what is wrong.

As a former Ground Search and Rescue member, I was trained to observe the environment for "spore"; signs that someone has passed through. Once I found someone, I was trained to observe their behaviour and their body for signs that they had been injured or that something else was wrong so I could stabilize them and transport them to a location where I could hand them off to a paramedic who was trained to use their equipment to make observations that I could never make in the forest.

When I worked in finance, I would observe the behaviour of my clients through their bank statements and by asking questions as we worked together to create a budget and a plan for their future. The observations would guide me in determining the tools needed to help them meet their goals. Which budgeting system, insurance products, and investments would help clients reach their goals.

Observations are important in all areas of life. Here are some observations that you can make around your property, and how they may affect your design.

1- Which way the winds blows most of the time as well as when it is the strongest like during a storm.
   a. Where to place windbreaks
   b. Where to place wind-pollinated plants like hazelnuts
   c. Where to place windows in a home or outbuilding to reduce cooling costs
2- What is the last place that has snow in the spring?
   a. These cold areas can sometimes be the best place for trees like apricots which boom early. Keeping them in a cooler area can keep them dormant longer so their blooms aren't killed by frost
   b. This area may be a bad spot for trees that aren't quite hardy enough for your cold winters
3- What does the rain do when it hits your land?
   a. If it sinks in quickly we can plan for well-draining soil
   b. If it pools you may have drainage issues, some plants may love being wet but it can kill others
   c. If it runs across your property bringing debris with it we may want to address potential

4- What animals enter your property? Where do they go when they are here?

    a. Included in or excluded these through your design

    b. If you want to keep deer out it is easier to do it by placing the fence on an angle other the perpendicular to the path of the deer, this encourages them to go around instead of jumping over the fence. To do this, we must know the primary approach angles.

    c. Heavy populations of some animals may encourage adding plants like thorny raspberries around your fruit trees to help protect them when they are young (in addition to trunk guards)

5- What rooms in your house get hot or cold, and when?

    a. Insulating your home can save energy, reducing your overall footprint on the planet

    b. You may benefit from a shade structure over your window that blocks the summer sun while allowing the winter sun in

    c. Tree placements can affect how the sun hits your house

    d. Tree and bush placement can direct more wind towards your house to help with summer cooling

6- Where are the ugliest parts of your property?

    a. A new home build may be best placed in the ugly spot so you can look at the beautiful areas

    b. These areas can receive extra attention for restoration and beautification

7- What sounds do you hear that you don't like and where do they come from?

    a. Trees and bushes, or even buildings like barns can be placed in locations where they can help block or dampen the sounds of a highway or a noise factory

    b. If you want something like a waterfall on your property, it can be placed so it masks the sounds when you are in your most commonly used areas, or the areas where you want to relax in most

Some of the observations may seem minor but can make a significant difference. There are other observations that we may make, for example, if you want to plant an orchard I may recommend digging a four-foot-deep hole and letting it sit for a while.

What can we learn from this hole?

> While digging the hole we will see what your soil is like. Is it sandy? Is it clay? Are there lots of rocks? Did the previous owners bury their garbage there?

> After waiting a little while we will see if the hole fills with water. Most fruit trees want at least 4' of soil before they hit the water table. There are ways to get around this but they should be done before planting your trees.

Watch nature and see what the land has to tell us. Some plants will tell us that you have lots of phosphorus in the soil and other plants that will tell me that a certain area is likely wet for most of the spring. There is so much to see.

Start practicing observing and making notes. It can be difficult at first, but you will get better with time and you will be rewarded for your diligence.

All your observations will save your designer time and save you money, and you will get a design that works better for you. This applies just as much if you are doing your own design work, possibly even more so, because you won't have a professional to guide you to the answers.

Zones

Chapter Three: Why Pay for a Quality Design?

You can make a design yourself; you may even be able to make a good design, it's pretty rare for someone to make a great design on their first project though.

Note that this chapter is about paying for a quality design, not all permaculture designers are worth their fees, just as not all electricians, mechanics, or teachers are. Not all designers will give you a quality design that is well-suited to your project.

It is important to find someone who will make a quality design.

The Designer

Picking a designer is a different conversation for another time, let's look at why you would want a good one. We'll look at my needs as an example.

If I were looking for a design for a brand-new property on raw land, this is what I would be looking for (keep in mind that I have especially complex plans):

- Permaculture Design Certification
  - Shows that they have committed to the study
  - They have been taught the ethics and design principles
  - They should have had a minimum of 72 hours of study
- Additional studies
  - Have they studied anything else to improve their design or observation skills like botany, environmental studies, pest management, soil life, etc
- Passive Solar House design skills
  - I don't want heating or cooling bills, a permaculture designer who can help design an efficient house is an excellent asset for my project
- Passive Solar Greenhouse design skills
  - These are a big part of my plans going forward. Someone who knows how to design Passive Solar Greenhouses and how to design a plan for using them well would be an asset
  - I want a series of ponds, a good designer would place them so they reflect the winter sun to the greenhouses if it works with the rest of the design. This could provide a few extra degrees of heat in the winter
- Rainwater harvesting
  - I plan to use rainwater for my drinking and for watering needs
  - A system needs to be designed correctly to be drinkable without additional filtration or treatment
  - The system needs to be sized correctly according to roof size, local rainfall, and water usage needs
  - Connects with the house design, designing a roof that is efficient to harvest from simplifies the system
  - This connects to the greenhouse design, some water storage may be included inside the greenhouse plus I would collect rainwater from the roof of the greenhouse
- Pond design and earthworks
  - I love water and water-based ecosystems and want interconnected ponds to observe and interact with
  - The ponds need to be built and sealed correctly
  - A waterfall needs to look and sound good

- Orchard design
    - A you-pick orchard to allow me to give away thousands of pounds of food every year needs to operate smoothly with minimal inputs while creating great yields
    - Pest management needs to be integrated into the design
    - Fertilization needs to be integrated into the design
    - Pollinator support needs to be integrated into the design
- Renewable energy
    - Off-grid with battery backup, this needs to be positioned on the property and designed in a way to meet my needs and be repairable

It is a complex design that has many moving parts. Having one person who has, at a minimum, a basic understanding of every part will allow me to make sure they are all connected into one big system.

There aren't many designers who can work in this many areas. The 5th World Permaculture team could likely do it, I could do it, and there are likely a few others here in Canada. You must be a real nerd and be extremely passionate about permaculture to spend that much time studying, the return on investment isn't high enough otherwise.

If you can't find someone with all of the skills, you may be able to find a team that will work together as members of one design firm. Having a design firm may be better than getting one person to do it all because you would get more eyes on your project.

If you can't find an individual who has all of the skills required and cannot find an existing team, you'll need to start picking people who specialize in each area. Hopefully, your Permaculture Designer also has a general knowledge to be able to connect all of the elements.

Characteristics of a Good Design

A good design may look simple but is the product of a great deal of thought and observation.

Designing in Zones

The zones of a permaculture design should be evident, with everything that you visit frequently being close to home, with the less frequently accessed areas being further away.

You won't run to the garden to grab herbs for that soup if you have to walk through 3 gates, past the barn, and through the bull's field to get to it. But you will use those herbs without hesitation if your garden is near the house, even more so if you have an herb garden just outside your kitchen door.

Permaculture design uses a series of zones which may be concentric circles. This is how we usually display it in educational diagrams, but in practice they vary in shape to meet our needs.

These zones are numbered from zero through five.

Zone Zero

> Zone zero is your home, the base for doing life. This is where you spend most of your downtime and is your home base for heading out to do activities around the property.

It is possible that an outbuilding on the property could be considered as a second Zone Zero if that is where you spend most of your time. An example of this could also be an outbuilding that is used as a home daycare, business, or market.

This is the zone that is managed most intensively (cooking, cleaning, etc)

Designing zone zero should receive just as much care as designing the rest of your design. How a house is constructed can have a massive environmental impact through energy usage while affecting enjoyment.

Most design projects contain an existing home. When doing a complete design I will several home upgrades that can improve efficiency and may recommend getting a home energy audit. If there is a big enough difference between the indoor and outdoor temperatures I will use thermal imaging to check for heat loss. I will check how the rain will travel from your roof to its final resting spot, check for leaky outlets, inefficient appliances, and improper seals around doors can be other elements of the design.

Another big opportunity for retrofits is to control how the sun hits your windows. Allowing winter sun in while blocking summer sun can reduce both your heating and cooling needs.

If you're designing a new home, your zone one's location may be affected by many factors including flood plains, fire threats, orientation, and views.

Zone zero is your home, this is essential for your enjoyment of daily life.

Zone One

Zone one holds anything that you need to access multiple times per day. For example an herb garden, social spaces like an outdoor kitchen, your kitchen garden, maybe even a few berry bushes or fruit trees that are for casual eating.

Zone one would be a good place for plant propagation beds to make sure your new cuttings get the attention they need during the sensitive initial phase of root growth.

Quiet and sensitive animals can also be in your zone one.

This zone is designed for efficient access from your zone zero. You don't want to have to walk to the far end of the property to pick some herbs for dinner or to sit outside for lunch. Your outdoor table should be close to the kitchen so it is convenient to eat there.

Zone one can be a circle around your zone zero but I find that it is often a patch near each door plus an area on one side of the building.

Zone one should be so easy to access that you can visit it without thinking about it. Trips to this zone are a natural part of your day.

Zone Two

Zone two is less accessed but may still be accessed daily. This can be where you plant a home orchard, something that gets mulched and maintained, possibly watered, but that you aren't in all day, every day.

This is also a great place for your storage garden. Your potatoes, winter squash, and other plants that are pretty much "plant and forget" go here. You may still water them but you won't be harvesting from here every day, most of your harvesting will happen in one big day.

Zone two is a great place for small ponds, possibly an outdoor seating area away from the house so you can relax. Your chickens can also live here, you can collect eggs and check on them daily, and you may even be able to watch them from the house.

This is all about convenient access without crowding everything right around your house.

Zone Three

Zone three is for commercial operations and long-term storage. If you are going to this space, you are doing it for a reason.

Large ponds for water storage, storage barns, large orchards, and market gardens; are some of what you'll find in this area.

Your market gardens in zone three could benefit from being placed near your zone two chickens so you can move the chicken bedding (and manure) into a compost pile and then into the garden efficiently.

Zone three may also house herds of livestock that are either on pasture or are for commercial purposes.

This area is about large-scale production and storage.

Zone Four

Zone four is almost wild but is cultivated for the foraging of foods and other resources like firewood. This would resemble the example of an "eight" or "nine" that you will read about in the "Sliding Scale of Sustainability Design" chapter. This area is wild but is curated for our benefit.

There may be more pond space in this area for water storage but it isn't a primary water source because of the distance.

You may go a month without visiting this area, it's almost a natural space but is maintained to serve essential functions.

Zone Five

Zone five is a natural, nearly unmanaged space. It may be used for recreation (some hiking or snowshoeing trails and forts built by the kids), and you may also do occasional foraging in this space.

This is a place to observe nature and learn about the patterns that we can apply in the other zones.

In an urban design, you may only have up to zone one, two, or three. If you live in a high-rise condo you will have zone zero, and you may have zone one on a balcony or in a grow tower inside your unit. You may not have a convenient zone 2 or 3 but may be able to get a growing plot near your building or even on the roof.

My residential lot was 50'x130' and it had three zones with zone one being bigger than usual because it was a shared space with my tenants. I had no zone four, but zone five was a forest that bordered the town. A space to forage and to learn about nature.

Designed With Your Needs in Mind

A good design isn't made off of a template; it may use patterns that are present in other designs but isn't a copy-and-paste situation because you aren't exactly the same as anyone else and your property isn't exactly the same as any other property.

If you lived in a row house and moved one unit over, it is likely that your design will change. This may be because there is a tree that changes how much sunlight you receive, or receive more water runoff, or the wind has been funneled directly towards this lot. This is a bit extreme of an example. You could probably keep much of the design the same but you may have also changed since the first design was completed and now have different priorities.

The design shouldn't be defined by what your designer is fond of. I love passive solar greenhouses but they rarely are a part of a project that I design. I love Earthships but have never included one in a design for anyone else because it was never the right fit for any of my clients.

Designers may want to include swales or the back to Eden method of gardening. They may want to include a 30,000L rainwater harvesting system or a commercial-sized worm compost system. These may be right for you but they may not be so you should consider every element.

Every designer has their own values and it can be hard to keep these from influencing our design work. A designer who is driven by a need for security may want you to go off-grid, while another designer who is motivated by frugality may want to do everything on the cheap while you would rather pay to get things set up quickly, correctly, and in an aesthetically pleasing way with minimal input on your part.

<u>Clearly Illustrated</u>

It is possible to have a good design without any drawings. This would have to be a very text-heavy design package to be able to explain what the plan is. If I was sending a text-only design to someone, it would be more of a prescription of systems and patterns to include than a typical design. The design would greatly benefit from even a few basic illustrations.

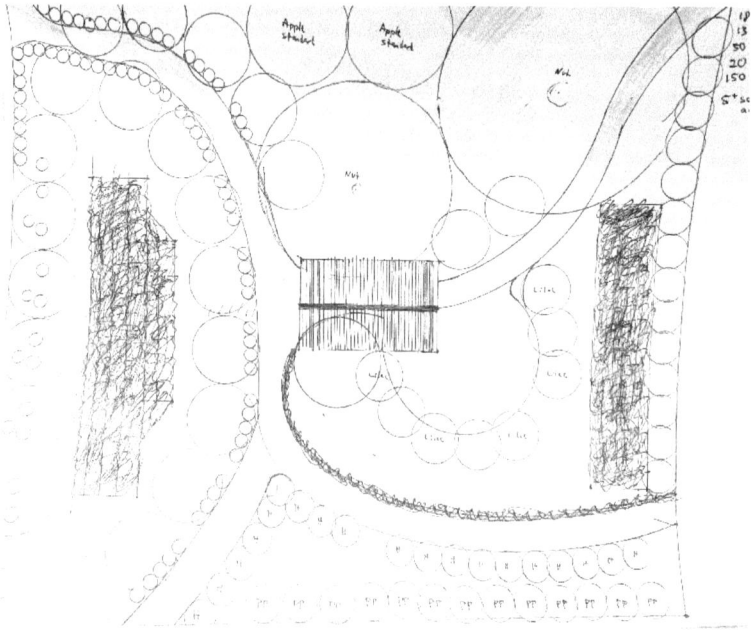

Back of the Napkin Illustration:

A back of the napkin illustration involves pencil and paper, maybe some colour but it is something that could be done while sitting at a café so you have a general idea of what is being explained.

You can see a very basic drawing on this page. This was a 2 a.m. sketch that I did to work through an idea for a municipal park that was set to be sold. I consider this to be rough work and would not be presented to a client as part of their official design. It is a great example of a basic illustration that gets the point across, especially when accompanied by text. This exampled included a lengthy description of what was shown.

For this drawing I used a few inexpensive architecture tools to help me do the curves and for all of the circles. They are just pieces of plastic that I can trace around, it takes the need for any artistic skills out of the project.

The "back of the napkin" style drawing shows you how to set out on your project and will give you enough information to start talking to an installation contractor. This person can dial in measurements and placements based on the ideas outlined by the Permaculture Designer.

The process of creating site drawings is not covered in most Permaculture Design Courses. This leaves it to the designer to decide whether they are going to stick with pen and paper or if they will learn a program like CAD or SketchUp. It's a big commitment to learn these programs.

These back-of-the-napkin drawings may be all that they ever learned to do. I have met some brilliant landscape designers who work on this level, generally doing the installation work so the lack of information isn't an issue. It's just a matter of trusting the process and trusting your designer.

I could work off of a drawing like this for my own projects with no problem because I know what it means. If I was just the hands-on guy and received these drawings I might get frustrated though because it lacks information.

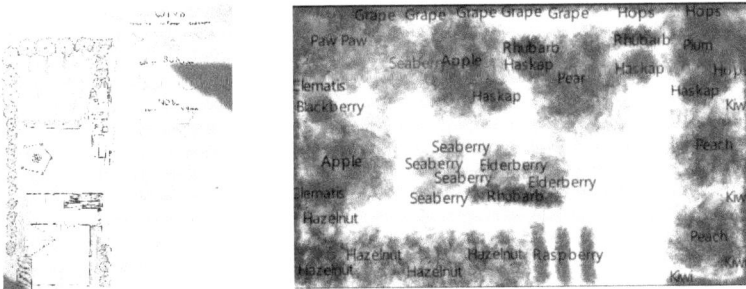

Hand Drawn Illustrations

Hand-drawn illustrations (including digital drawings) can be beautiful and effective. They may not offer much more information than the napkin drawing but are more likely to be drawn to scale which can be helpful. This has enough information that I could use an architectural ruler and the scale reference on the drawing to be able to layout and install this project even if I knew nothing about permaculture design.

Above-Left is an example of the first design I ever did. This is a simple drawing that I added some colour to so it would look better when I was getting graded for my permaculture designer certification.

This was before I learned about landscape design for beauty and involved working around having shared spaces for tenants. It was accompanied by plant lists, planting information, and basic rainwater harvesting information (including a parts list for the rainwater harvesting system). It was functional for what was needed at the time and is the design that I used for this property with some modifications because I planned to move from this site.

These drawings are all that most projects require, especially if the location is fairly flat or evenly sloped, without lots of hills or rock outcroppings.

To the right of the first drawing an example of a quick but more artistic digital drawing that I created as part of an Instagram post showing how much food could be produced in a 30x50' lot with a "wild feel". This drawing features many colours and textures to represent the various plants in the space.

The hand-drawn illustration is what most of my clients will receive. It will usually be done digitally so I can make revisions more easily and can include a version that contains any critical measurements.

We can take it a step further in a few ways, let's take a look at these more advanced options so you know what is available.

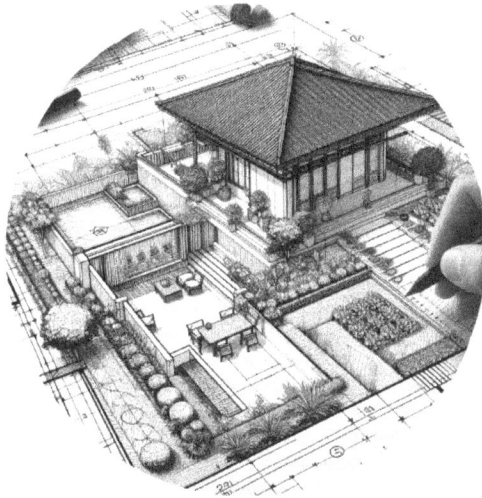

Blueprints or Architectural Drawings

Any project that involves a lot of hardscaping (stone pathways, stone stairs, retaining walls, etc.) requires specific measurements so the correct materials can be ordered in the correct quantities. This is usually completed using digital illustrations but can be done using overlays for a paper drawing.

For walkways and simple stairs, it allows the installer to know what is needed to install the project. Every measurement is available in the drawing or can be calculated easily.

As for retaining walls, this provides information on the shape and size of the retaining wall.

Architectural drawing is a skill set that should be accompanied by hands-on experience. An average designer will not know that the local company builds their retaining wall blocks in lengths of only 24" or 48"; on the other hand they have blocks that are 24", 27.5", 46.5", 48", and 51.5" wide.

Without understanding how these blocks are designed, you may end up with drawings that must be changed or that require cutting 6" off the end of a truck-load of 2400 pound blocks to make them fit. This wastes time and materials and may end up looking worse than making the wall 6" longer.

The first time I included a retaining wall in a project was while helping a friend renovate his Dad's house. We were replacing an existing retaining wall that was failing. We had the manufacturer come out and tell us what was possible. They also made the drawings for us.

I participated in the installation so I could see what did and didn't work on location. Unfortunately, the company we used was chosen because it was the landowner's childhood friend, not because they were good at their job. After the 5th major revision to the drawings over a period spanning 14 months (because they had missed entire sections, and mis-measured by up to 4' in some areas), I ended up creating the rest of the design myself with their engineers approving it at the end of the project.

This involved risk because there could have been an issue that required us to tear it out and start again. We had already been delayed by a full year waiting for new revisions and we were nowhere close to a workable drawing, so we decided to press on while knowing the risks. The finishes on the blocks also had bubbles and damaged parts on arrival, these would hold water which would freeze and result in premature failure.

Choose your suppliers wisely, just as you choose your Permaculture Designer wisely.

If making architectural drawings now, I would create a drawing that I think works, send it to a reputable installation company near the client, and have them let me know if it makes sense with the standard material sizes they have access to.

I would leave the technical parts of the drawings to the architect but include the outside measurements, stair widths, path widths, and total elevation changes. The design may also include a buffer zone within which certain elements can be moved to accommodate changes during the build if the client chooses a different manufacturer or if the installers run into an issue that requires a modification to the design.

This would mean that my drawings wouldn't be true architectural drawings but would be close, while having blueprints accompanying them.

As you can see, these technical drawings are great for the installer if done correctly. They are a whole different game when compared to what most Permaculture Designers work with, even the people who specialize in these drawings can get it wrong and cause major headaches.

3D Renderings

These are 3D models of the entire property or of specific areas. These take a lot of time and the skill of working with the 3D design software.

A 3D rendering of a flat location can be fairly simple, creating a flat plane, importing plants, and modeling some really basic buildings. The goal can be to show just enough to help you picture the design in your own space.

I chose to do this when putting an addition on my parents' house so my Nanny (grandmother) could envision the space she would be living in. She could not see the space in her mind while looking at the 2D drawings, most people cannot do this. Because of this, I created a 3D model and digitally walked her through it as we physically walked around the space.

This can be a very valuable tool if one of your major values (which we will explore in the chapter "Guided by Values") is beauty so you want something closer to edible landscaping than a wild growing space and want to see what it will look like when completed.

These drawings are valuable for complex landscapes as well. A designer who is skilled with their modelling software can import topographic data into the design to replicate the hills and valleys. The designer can use this to add another layer of complexity to the design.

This complexity may allow the designer to work with the elevations so the beautiful red maple tree on the far hill is framed perfectly by an opening through the fence. This will be seen while sitting at the bench across from the birdfeeder.

The topographic data may also allow the designer to create a series of interconnected ponds that will flow into one another as the rain fills each one to the spillway height. The water flows over the spillway into a swale or ditch that moves the water down to the next pond. This interconnected pond system could be for water storage or it could be to build habitat.

The interconnected pond system could also be part of a fish breeding program in which each pond is emptied into the pond below as the fish reach the next phase in their growth cycle.

Being able to use accurate LiDAR data to map the land would give a high level of confidence in the viability of the fish farm design, which would be difficult to achieve otherwise.

I have done a lot of work with 3D modelling software. Only recently have I began exploring the intricacies of including topography. I look forward to working with it more, but for now would likely create a 2D drawing while referencing the LiDAR information to ensure accuracy. Including some perspective drawings, or a rough 3D rendering, not detailed enough to be usable as a technical reference.

Perspective Drawings

Perspective drawings are an artist's rendering of what your property may look like. If your designer has created a 3D model they can capture images of it from different places within the model. Without the model, they are usually hand drawn and require yet another skill set in the arts.

These drawings are what you might see when a building project is underway and a sign is installed with a picture of what it will look like including landscaping.

These are often included if you have 3D renderings created. A hand-drawn perspective drawing is a service that typically comes with a hefty price tag and is reserved for high-end projects. These often involve hiring artists if a digital rendering isn't part of the design process.

I would also include perspective drawings if I was making a proposal for a project in a public space. The drawings would get more attention and may gain more public interest. Sometimes a "back of the napkin" perspective drawing can be a helpful tool as well to express a general idea.

Summary

There are a large variety of illustrations and all of them can be effective depending on your needs. I may even offer a "Back of the napkin" drawing to someone if their budget is tight and they only paid me to walk through their property with them to pick my brain.

If you expect 3D modeling, be prepared to pay a premium and to search for a Permaculture Designer who has this skill set. This will limit your options significantly.

If you are expecting to give the plans to a landscape contractor and have them install it accurately without additional guidance, you will want something closer to architectural drawings.

If you want to have organic design (the ability to make intuitive changes) happening throughout the installation processs, you can get away with the simple napkin drawing. Every option is good when done well, but every option can also be done poorly. Anything better than a napkin drawing should be drawn to scale.

A quality design doesn't need to be fine art. It does need to incorporate the minimum amount of information that you require to proceed. It also needs to be incorporated in a way that you can understand.

A note on working with new designers

If your designer is new to the trade and hasn't learned how to create drawings yet, they could create a general layout of the property (boundaries, buildings, other existing major features) and then make a collage of what is in each area. This should be accompanied by a thorough written description. The drawing may also be labelled with numbers or a grid that is used to help understand locations.

They can also take satellite images from a service like Google Earth Pro and sketch ideas over them.

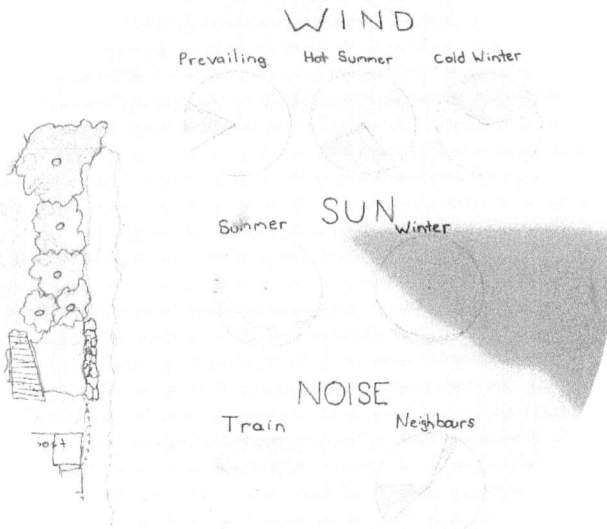

WIND

Prevailing   Hot Summer   Cold Winter

Summer   SUN   Winter

NOISE

Train   Neighbours

Soft

Sector Analysis

Sectors in your design are a visual representation of significant environmental factors. Every Permaculture Designer should include these in their designs. It doesn't take long to draw them out and it is information that the designer should collect before starting their design work.

Sectors can show:

- Where the Summer sun rises and sets
- Where the Winter sun rises and sets
- Prevailing wind direction
- Directions of noises
- Water runoff entering your property
- Common wildlife traffic
- Wildfire risk
- Human traffic passing through the area

Not all of these will be relevant for every design but wind and sun are the minimum that should be included. This shows that the designer considered these factors in their design work, not installing a sun-loving plant on the shade-side of a building or make other similar mistakes. You can see a simple example of a sector analysis on the previous page, this came from my first design shown in the "Hand Drawn Illustrations" section above.

Your sector analysis doesn't need to be complex, but if your designer can't tell you what the sectors are, you are at a greater risk of having a design that will not meet your needs.

<u>Contains Important Information</u>
Putting all of the required information on a drawing isn't practical. Accompanying text should include some of the following (depending on what you agree to when hiring your designer):

- A list of plants to include
- Important measurements like tree spacing
- Material lists for hardscaping
- Contact information for contractors or specialists that have been referred.
- Soil information
- Notes on Specific areas or projects
- Explanations for important elements (like a deer hedge which must be installed properly)
- Detailed information for hardscaping
- Specifications for ponds, rainwater harvesting systems, etc
- Safety warnings about dangerous plants observed on the property or included in the design; about maximum slope angles for ponds, etc
- Recommended insulation or window specs
- Notes on recommended renovations, etc
- Recommended mulches, walkway materials, etc

This is not an exhaustive list as many projects may include other information, yet most will exclude most of what is listed here because it is outside of the scope of the project. The purpose of this section is to let you know that there should be additional information which enables you to use the design.

Chapter Four: Guided by Values

A look at the impacts of our values

We all have values that guide us but if they aren't clearly defined, they cannot guide us well and we may even do things that work against our values. Taking the time to identify your values may or may not change our designs but it may, and may also change the entire direction of your life.

My own values are defined well by the ethics of permaculture: "Earth Care, People Care, Fair Share".

I want to take care of the Earth that we have been gifted. It is beautiful and has been the focus of my fascination since childhood. It has also been a source of both awe and wonder. This encourages me to work with nature to restore the Earth. The awe and wonder are significant factors in my decade-long journey of studying permaculture design, ecological construction methods, rainwater harvesting, passive solar greenhouse design, retrofitting homes with energy-efficient upgrades, and more.

I love people and it is my goal to love like Jesus did; this isn't a religious book, that is just the phrase I use to explain how I want to love people from all walks of life. This guides me to create a plan for my life that allows me to help others. Sometimes helping people looks like volunteering at a local café that gives away food. Sometimes it means letting a recovering addict sleep on my couch, or defending a local greenspace.

This value has deep connections with the third ethic of permaculture; Fair share. This is about sharing the surplus with the earth and with other people.

Fair Share defines how I want to run my own orchard as a you pick that allows people to exchange a little bit of labour for food. They can pick a basket of food for me, a basket for the food bank, and a basket for themselves. This system lets me give food to the people who can help themselves by trading labour for food it also lets me help people who cannot come pick food for whatever reason.

I would not have been able to refine my vision this well if I hadn't taken the time to identify what is really important to me. This platform will define many things within my design including the layout of buildings to accommodate the processing of food for the food bank.

This results in many other changes in the design, things that would not need to be included in a purely for-profit project.

What are the values that guide you and how can we figure them out?

Understanding your values may be intrinsic for you but for others, it involves work that goes well beyond writing a couple notes on a worksheet. I spent years trying different things, experimenting, journalling, and working with my therapist.

My years of work can't be replaced by a quick exercise, but I have found a few exercises and questions that can get you into the general ballpark. My recommendation is to work through these exercises and then continue to think about it for the rest of your life. There are few things more important than knowing what is important to you.

Let's take a look at a few techniques, don't rush by this. This may be the most important part of this whole book.

Ask a friend

The first exercise requires the least self-knowledge but does require a vulnerable conversation with someone who really knows you.

Your friends see what you do, they hear how you talk, and they know the patterns of your life. Sometimes this can provide them with some insight which can help you understand yourself.

This is an exercise that you should only do fully with someone that you have a trusting relationship with because it can lead to a sensitive conversation. Doing this with someone you don't trust could lead to some unpleasant conversations.

You will set the situation up by letting your person know that you want honest answers even if they are unpleasant or take a while to come up with. Then you can choose how much of the backstory to share with them before asking the question.

The question is: *"Based on what you know about me and what you have seen, what do you think my core values are?"*

This question is a big one and it may take some time for them to answer it. It is a vulnerable question because you are asking someone to identify who you are at the core and to put it into words regardless of how it may make you feel.

I have a friend who I could just walk up to and ask this question without setting it up at all. He would take a moment to think and then would respond. He would ask what it's for after he has responded and we might chat a little longer, he may even ask me the same question.

He would then reach out again sometime in the future, maybe a couple of hours later, maybe a couple of days later. At this point, he would then add to his answer after having thought about it some more.

I have another friend to whom I would explain the whole purpose of the exercise because this is what they would need to answer the question.

If you don't have at least a small idea of how much back story to give, you may want to either think of someone else because you want someone who knows you or ask multiple people to get a bigger sample size for your data.

Note:

If you don't have anyone you trust to this level, you can ask it on a more surface level to others that you have spent time with but you'll want to get a bigger sample size by asking more people. The same question can be asked in a casual conversation. You can adjust the question but it is worded very intentionally so I recommend sticking with the original wording or to something close to it.

These responses from more casual acquaintances may be based only on your behaviour at work or on the soccer pitch because that is the only context in which they know you. This lack of depth isn't ideal, but it can still help you get a new perspective on what your actions say to others.

Asking people from different areas in your life will help you get a more complete picture, especially when asking this second group.

While writing this book I paused to ask a few new people this question and my answers were the following:

> "*Sustainability – Trust – Human connection – Adventure – Nature – Growth – Innovation – Faith – Freedom – Respect.*"

> "*I'd argue sustainability, kindness and compassion, active communication in both the active listening category and the actual discussion aspect, I'd also argue integrity and honesty. You might not demand the last one from people but unintentionally you bring it out in people*"

> "*Giving. Giving to people, to the planet, and Jesus*"

As you can see, many of these could easily guide me to my 3 core values of Earth Care, People Care, and Fair Share, there are also hints of other things.

Note:

I had to pause after reading these, it felt so vulnerable
for people to explain me so well in casual conversation.
In these moments I felt vulnerable, exposed, and seen,
but also safe because I asked people whom I trusted
deeply. I didn't expect such positivity or such flattering
responses. I asked people who I knew were willing to speak
the hard truths as well.

As we pick them apart for themes we see:

-Sustainability was directly mentioned, this is Earth Care
-Kindness, trust, connection, respect, compassion, kindness, and honesty all connect back to People Care
-Fair share wasn't really mentioned here but it stems from People Care and Earth Care, it is more of a result of the other two

One person mentioned that my words make it seem like money is important to me but that my actions show that it doesn't matter at all.

This person is the one with whom I talk through dreams and business ideas. As the Excel nerds that we are, we run the numbers so we can understand the financial implications to see if something is viable as a hobby or as a business idea.

He also sees that I don't spend money because I don't care enough about money to commit heavily to earn it. This is why I mostly make my living by doing things that I love instead of doing what will earn me the most money. I view money as a tool to build the things that will help me love people and the Earth. I believe that the only reason to have a massive amount of it is to be able to be outrageously generous (Fair Share).

Why why why why why?

This second exercise is one that you can do alone but it will take more introspection than the first exercise. This is a much more common practice that is taught in countless seminars and books.

To do it we need to get in touch with our inner two-year-old, the part of us that asks why over and over and over again until we cry. We may not need to actually cry, but we want to get to the point where it moves something deep inside us.

The purpose is to start with a surface-level desire and to understand what the root of that desire is then.

We start with a simple question like "Why do I want a food forest?" or "Why do I want to homestead?" or even "Why do I want to start this business?"

I did this exercise with Stefan Sobkowiak during an orchard design masterclass when he asked "Why do you want an orchard?" so I'll walk you through something similar to what my answers looked like (I was silly and didn't record them, please record yours!).

Why do you want an orchard?
> I want to grow a lot of food.

Why do you want to grow a lot of food?
> I want to be able to give food away to people who need it.

Why do you want to give away food?
> Because I want to make sure people aren't hungry

Why you do want to make sure people aren't hungry?
> Because no one should go hungry.

Why do you want to grow food so that no one goes hungry?
> Because I am called by God to feed the hungry.

Why are you called to feed the hungry?
> Because they are all children of God and I love them.

Why do you want an orchard?
> Because I love all of God's children and believe that I am called to ensure that they don't suffer hunger in a world filled with food

At the end of this exercise, I returned to the original question to see how my answer had changed. Because of this exercise, I know that my driving reason for wanting an orchard is because I believe that every person is a child of God and that they shouldn't suffer from hunger. I believe that it is my purpose to feed these people.

That is a powerful motivation for me.

This motivation will do two significant things:

1- It will keep me going when times get tough.
2- It will provide a guiding factor as I choose what plants to include, how I will harvest, how I will market, etc.

Finding values through reflection:
This third exercise takes the most work, the most honesty, and requires some very clear memories. It isn't the most reliable tool for identifying your values but it is one more tool in your toolbelt to complete this important task.

They say that hindsight is 20/20 but that's only true if you take the time to look back. Otherwise, it is something that you may as well be blind to.

Journalling is an excellent tool for helping us look back at our lives, you don't need to have journaled all your life to benefit from it now though.

I want you to recall some memories. Some of the best times in your life when you were truly happy. Times when you were proud of yourself. Include moments from childhood if you remember them. Write about it, even if it's just a few sentences.

"They say that hindsight is 20/20 but that's only true if you take the time to look back."

The next step is for the current day you to write a letter to that version of you. Talk about that moment and explain it to your younger self. Talk about what you were feeling in the moment and what you feel now. Write about why you are still proud of that moment. What does that moment tell you about yourself?

I remember one time when I quit a job, I'll use that as an example for you here and then I'll make up an example for the other option because I can't think of an example that is appropriate to share in this context but that also taught a valuable lesson.

Journal Entry:

> I quit my job today and I'm scared but I'm really glad I did it, I'm proud of myself for how I handled it!
>
> I showed up to work early and was waiting in my car, watching a YouTube video about something unimportant before getting to work. My boss knocked on my window, so I rolled it down and he started yelling at me. I have seen him getting angry at other people lately but this time it was directed at me, and I recognized that I didn't deserve it at all.
>
> I let him finish his tirade and then I calmly explained that he hired me to do carpentry and that I was never trained to fix tractors or snowplows. After that, I asked, "Do you want me to go home, or do you want me to get something done?".
>
> He said it was up to me, so I got to work, I worked hard and fast while thinking about praying about how to react well. I had a nice long drive towards the end of the to-do list and it helped me understand that I shouldn't mention the yelling because it wouldn't be helpful, but that it was time to move on from this job.
>
> He left for the day, so I left him a note. I started by giving an update on all the work that I had done as well as mentioning a part that still needed to be replaced once he sourced one. I followed that by explaining that I had other work that I could do so I didn't need his work anymore and that I'd help out from time to time if he really needed me but not to make work for me.
>
> On the drive home, I was terrified because I had just said bye to my most reliable source of income, but I also was really happy with the decision and with how I went about it. I better get to figuring out what's next, I need to double down on Eden's Refuge and should take a side job during the transition, fortunately, I think I was just offered a good fit.

The Letter from current me:

Mate, that was a big day! You were feeling a whole lot more than what you wrote there. You were unsettled and feeling insecure about your financial situation but you were also so excited for what the future held. You were upset at being yelled at for something that seemed entirely unfair and you wanted to say something that is not suitable for a book that your Grandma will read then drive off.

I'm so proud of you for how you took the time to think before responding. Only a few months before this, you would have put the car in gear and left him in the parking lot then sent a not-so-pleasant "breakup text" later on but you didn't do that! You took the time to listen, you looked at him and recognized that he was experiencing a lot of emotions and that there must have been something else going on and you responded appropriately.

You used your love for people to give you patience and strength. You used that same love to give the angry man a chance to calm down. You even chose not to poke at his sore point or bad behaviour in your letter of resignation. You loved like Jesus in that moment by not allowing someone (yourself) to be treated poorly and without attacking the offender.

You have shown me what is actually important to me as I look back on this. You established boundaries to take care of yourself and you treated others with love no matter what was happening.

You may see now how this is a difficult process to pick apart sometimes. We are trying to identify the motivation. In this case I wanted to identify why I felt proud of how I reacted. I know that I was proud of not losing my cool but there was more to it than that.

Next, I want you to do the same thing with a time when you did something that you weren't proud of, maybe even something that you feel shame about. And then we're going to reflect on why you felt that way. Let that mistake teach you a lesson. Here is my example based off of an all-too-common story we see with kids:

Journal Entry:

I don't know why I keep spending time with Bob and Gary. I don't like them at all, and I do things that I don't like when I am with them. I want to be with better people, but I don't know who will like me in my class. I don't know why I keep going back to them.

Letter from current me:

Hey buddy,

I remember that time, 7th grade was tough. I remember how conflicted you felt. You felt so bad when you were with them but you couldn't leave them. You said some pretty mean things to people. You didn't label it as bullying at the time, but looking back, I know that you felt that it was true. That is why it bothered you so much.

I have been thinking about it and I finally understand why you kept going back. You were so scared to leave them because you didn't know who else would like you.

You needed to feel secure, and it was such a powerful need that you were willing to be around anyone so long as you thought they wouldn't leave you. You hurt people to feel like you belonged.

I promise you that we will meet some of the most amazing people later in life and that I have some of the best people in my life. I still have this need for security, I deal with it differently though, now it shows itself through how I set up solar panels so I have control of my electricity and in that I have a massive pantry lined with jar after jar of nutritious food from the garden so we will never go hungry even if the supply chains get disrupted.

We have security now and we are creating more security, but we are doing it without sacrificing our comfort and without doing things that we don't feel good about.

Step back from that group, it will be hard but you will find others who will let you be someone who we can both be proud of.

Thanks for going through this hard time to teach us an important lesson.

These two examples came from two very different angles to show the same value, that I love people. One was through understanding why I was proud of my younger self; the other was through understanding why I felt bad about something that I did.

The second example also showed a need for security. In my letter to me I only recognized the general need for general security but didn't recognize that I was looking specifically for security in relationships and in community. Reflecting on this practice later on would show me how close I came to getting to the heart of it.

Your examples may look very different. You may find that your values are specific to something like human trafficking or hunger or that you believe that seniors deserve a better quality of life than they are getting. The important thing is that you get to the heart of these events.

How do these values affect our designs?

These three exercises will hopefully have helped you confirm your values. I used all three of them and then visited the results from all of them a few days later to see what else I could learn.

Understanding these values changed me and I now have "Earth Care, People Care, Fair Share" mounted on my wall with their definitions as I listed them in the chapter "What is Permaculture". These shape most of my decisions (I'm still human and make mistakes).

There are many values in the world and each one will affect plans differently. You may value profit, the environment, and family; or may value beauty, your friends, and victims of human trafficking. You may value nature and human connection, or you may believe that the spiritual journey is the most important part of everything you do.

Ask your questions, listen to the answers, and then sit with them for a while as you reflect on your life through the new lens that has been given to you. See if it is true. See if it is accurate, were your friends close to the truth? Were they completely off the mark? Be honest with yourself. One of my friends was so accurate that I took a few hours to recover enough to resume writing. For those of you who have read the Inheritance Cycle (a series of books written by Christopher Paolini, beginning with the book titled Eragon), this felt close to the experience of hearing one's True Name.

Now that you have, at a minimum, a rough idea of your values I want to look at why we did this. I believe that this is so important that I have considered refusing to do any major design projects for people who haven't defined their values.

I would help design a small home orchard or a rainwater harvesting system because these are small projects that can be adjusted if needed. On the other hand, I probably wouldn't design a multi-acre orchard, a homestead a retreat center, or a sustainable business without defined values.

There's too much risk for mistakes that would have too many zeros in the total cost. Too much risk that my design will set them back by years. Too much risk that I would do more harm than good just because they didn't actually know what they wanted.

The table below shows how some of these values may affect specific elements of your design.

The Effects of Values on Design Elements			
	Food Forest	**Solar Power**	**Greenhouse**
Security	Prioritize crops that can be stored well, maybe include some for creating biofuels. Include lessons on creating more plants.	Oversized system with large battery bank, generator backup, and lessons on maintaining your system.	Passive solar greenhouse with an off-grid, automatic backup heating and cooling system. Include aquaponics as a protein source.
Helping People	Prioritize crops that will help the food bank the most instead of focusing on the ones that you like most.	Additional low-power inverters and a few removable batteries to bring to neighbours during outages. Grow food in the shade under a solar array to give away.	Create a space where people can experience a beautiful green space during the cold, dark winter months. Plan yields to give away. Spare room to house people.
Caring for Seniors	Talk to a nursing home to see what foods the residents may	Racking for the array would be designed as a shade structure to	Wheelchair-accessible tours, grow food care packages for

	want (apple sauce, prunes, etc). Maybe partner with the nursing home to create a food forest at their nursing home with birdhouses and bird feeders where the residents can enjoy them.	provide a resting place for your elderly neighbours who walk on hot days. Small water mister to help them cool down even further.	seniors, create a community where seniors can share their knowledge with us (give them a sense of purpose)
Education	Cater to hosting schools and adults for educational trips. Limit nuts (allergens) include a shade structure for lunches and discussions and pay extra attention to hazards like roads.	Make all elements visible for workshops or tours. Include a small module and alligator clips to directly power lights or other small gadgets to allow for experiments and lessons.	Add a classroom to a passive solar greenhouse so you can teach classes year-round. You may also add a commercial kitchen to teach cooking.
Environment	Include more plants that support local pollinators even if they are less productive. Water management is prioritized. Include an area specifically for wildlife support.	Selecting batteries that are more environmentally friendly even if they use a lot more space or are more expensive.	Include rainwater harvesting, and use the most environmentally friendly insulation. Use recycled materials where possible.
Beauty	Curved pathways with features like statues or a pond with a waterfall. Seating areas designed for enjoyment. Special attention	Change the placement of the solar array or create more attractive racking solutions even if it's less efficient. Add a beautiful visual	Choose exterior finishes to compliment the environment. Include a water feature or exotic trees. Adds screens over all openings

	to bloom times and colours of trees and flowers. Special attention to ground cover between rows.	statement that distracts from the solar array.	and release budgies into the space, add habitat for them. Add feature lighting for beauty

As you can see, some of the values make small changes, while others can have big implications (like avoiding nuts in a food forest designed for school tours).

Some values like long-term financial gain may mean designing for efficiency of harvesting or putting more money into the installation to obtain yields sooner. The financial emphasis suggests doing significant market research before starting to design at all.

Other values may prioritize space for kids and dogs to play, maybe even including a rock-climbing wall on the north wall of the greenhouse or including a sandbox or a slide in that same greenhouse so you can spend time with the kids "outside" in the winter while still working in your garden.

I hope that you are starting to understand why it is important to understand your values before designing your property. Asking your designer for a food forest is great but as you can see it could be designed in straight rows or with organic curves or in a more naturally organized chaos of "guilds" in clumps.

I studied Stefan Sobkowiak's work which is a food forest designed into a commercial orchard so my mind may go to that style of design but you may have looked at examples on YouTube from Canadian Permaculture Legacy or Edible Acres where they have a much more organic design style which is less "commercially viable" but, having toured the Canadian Permaculture Legacy property, I can attest to its beauty and that it produces abundantly for personal use as well as for smaller scale commercial production.

These values are also important because your life may change over time. I am a single man right now but I have dreamt of being a loving husband and father since I was 7. It would be foolish for me to design a house that doesn't accommodate that potential future that I value so deeply.

This may not mean building a big house, but it could mean designing a big house. A big house that can be built in phases depending on the direction that my life takes. This approach is much better than designing a small house now and adding a series of additions later without forethought.

Planning ahead ensures that this part of the house will flow well into the future build instead of being a sectioned-off series of additions which are built without any real planning or consideration for the next phases of the build.

I have been in houses that were added to repeatedly and the additions are usually awkward, clumsy, definitely ugly, and usually inefficient.

I have examined properties with designs that didn't extend beyond the next project. This often resulted in things being positioned inconveniently. People didn't end up going to the compost pile that was behind the barn, or the chicken coop that was just a little too far from the house. The failure to design well resulted in a failure of the design.

Take the time to do these exercises and to revisit them regularly along the journey. When you get your first design from your designer, I recommend that you come back to these values to make sure that it works with what you want now and in the future.

Chapter Five: The Sliding Scale of Sustainability and Regeneration

As a Permaculture Designer, my clients usually like the words "sustainable" and "regenerative. These are not clearly defined terms though and they will become more confusing as big companies use them in their marketing either legitimately or through greenwashing.

We know intrinsically when something is destructive. Something inside us hurts when we see a massive oil spill, a logging operation clear-cut every tree in an old-growth forest, or the aftermath of war.

We know when something is regenerative when we see examples like Geoff Lawton's "regreening the desert" project in Jordan where they restored part of a desert into a lush landscape. (Lawton, 2016)

What exists between these two extremes? Most situations, likely including your own, will be in the vague area between these extremes of destruction and restoration.

I like to use a one-ten scale with clients to define how sustainable they want to be. Ranging from "doing a little less harm" to "living as a part of the land". Let's look at a few examples of the sliding scale in different contexts.

Examples of the Sliding Scale of Sustainability		
	Cash Crop Farm	Forestry	Market Garden
1	Using cover crops to reduce erosion and use of fertilizers.	Assessing a site so you can protect small ecologically sensitive patches.	Asking customers to bring their own containers.
5	Using no-till methods, planting cash crops between rows of trees and bushes that act as a windbreak, fertilize your crops, support wildlife, and provide additional yields	Strategic harvests of less than 30% of the trees, starting with the least healthy. Leaving the understory intact and replanting with trees suited to the area. Surveying the land to protect sensitive areas. Actively working to clean up ecologically damaging plants. Possible use of coppicing. Removal of any tree should benefit the ecosystem.	Focus on perennial systems. Using birds, wasps, etc for pest control. No-tillage. Watering with rainwater. Providing habitat for pollinators, birds, etc.
10	Foraging in a carefully supported ecosystem	Walking into a forest and removing some standing dead trees or trees that are growing too close to each other. Carrying them out without machinery to avoid damaging the forest.	Create an ecosystem that you can harvest from while leaving a surplus for nature, including many plants that are there only to support local wildlife. Only sell to people who live within walking or biking distance to reduce the carbon footprint.

There isn't any space on this scale for continuing in destruction, but it also isn't an elitist scale. It starts with "doing just a little bit better". Every number on this scale is good, every number is a step forward.

Let's look at some of the impacts of this scale.

Financial Effects of the Sliding Scale

We may think that we want to be a ten, but when I look at my own dream property design, I'm about an eight at best. I'll include a large orchard which would operate at around a seven on the scale and then would plant out another section that would be a ten.

The ten wouldn't provide many monetary yields. It takes too much work to harvest that ecologically sustainably for it to be financially sustainable. I would spend time working in the space to help it be healthy, similar to how the original inhabitants of North America took care of our forests so well that explorers called it untouched land that was lush with life.

I'd relish the idea of living a life that is a ten on this scale except that I'd also hate it. No computers, no cars, no mangos shipped in from a warm climate.

Fortunately, I don't need to live a ten to make a big positive difference. Even the Regreening the Desert project would be around an eight. The initial installation was certainly a ten, it is an amazing project that transformed that space from death to life, but the continued operation is probably closer to a six to eight because it focuses on meeting the needs of people.

From an economic perspective, we can make a good profit at most points on this scale. Here's a very brief look at a few of the impacts at various levels of sustainability in a cash crop situation:

1- Reduces chemical inputs and improves quality, raising the value of crops over conventionally grown crops.

5- Increased diversity of yields, increased resilience, and adding some high-value crops make up for the reduced area dedicated to the cash crops. Total financial gain is generally higher than at level 1.

10- Low yields of high value, crops. Every bit of produce is the equivalent of a hand-crafted garment in comparison to a slave-made one from a mega-store. This is a massively resilient system that would continue to thrive without human intervention. High labour costs which can reduce net financial yields substantially, possibly to a point below being economically viable.

Environmental Effects of the Sliding Scale

The reason many of us want to be a 10 on the scale is because of the environmental effects. The scale, of course, is organized from least environmental gain to the greatest which makes this section easy to understand and a little more difficult to write, yet it would be silly to not mention the main reason for wanting to be a 10.

We'll look at this from 10 to 1 this time, looking at the context of an orchard. In this example, we are starting with a vacant field. These are just examples and are not templates to be followed or a way to categorize how sustainable a property is. Every design is unique and I am just one designer creating one fictional description of one fictional plot of land.

Ten

We are creating something that prioritizes environmental needs over human needs.

This means that environmental yields come first, to the point of neglecting the importance of all other yields. We will plant a dense forest (in my local context that is what nature wants to create, yours may be different). We will plant with a massive diversity, including everything from early succession bushes to massive slow-growing oaks and maples. We will also plant groundcover, flowers, herbs, and everything that is well-suited to that space.

We will speed up the natural process of transitioning from bare field through succession to a mature forest by making sure it always has what it needs.

An ecologically pristine system will produce yields that we can obtain but those aren't factored into the design, they are just happy coincidences that we can benefit from as we forage from this space. It will yield nuts, berries, herbs, and more but you will have to explore the forest to find what you are looking for.

Nine

We start to consider ourselves a bit, we want to obtain a yield whether it is financial, food, or beauty. Our needs are still placed well behind the priorities of ecological benefit but we can make a few intentional selections.

Instead of using fully native versions of plants like black elderberry, we may start to include cultivars that produce berries that are more desirable as a product.

We can choose a grafted black walnut tree that produces better nuts than a seedling might, this is beneficial both for us and for nature. We could even include some apple trees that were grafted onto a full-size rootstock like Antonovka

These better cultivars would still be planted in a wild design but we can start to include narrow footpaths for easier access to the crops. These will be some of the same paths that nature ends up using.

Eight

Now we are starting to obtain a significant yield. We are entering the group of numbers on the scale that aims to balance personal gains with environmental gains fairly evenly, this ranges from about three to eight which is where most projects will exist.

At eight we are still leaning heavily towards environmental gains over human gains but are looking to increase our yield compared to number nine.

We have already started to shift towards the cultivars and varieties of plants that will be better suited to an orchard setting. These Cultivars and varieties have many benefits like greater yields, better flavour or storage characteristics, or disease resistance.

We will look at how we arrange the trees. Instead of creating a forest, we can start to create pockets of trees with paths between them. These wandering paths increase the amount of forest edge when compared with straight rows. To quote the regenerative band Formidable Vegetable:

"The edge is where it's at
And it's the most productive place
And my mama always said
If you're not living on the edge
Then you're taking up too much space" (Mgee, 2013)

The edge is where there is the most production and biodiversity, it also provides more opportunities for wildlife to safely cross a path without being disrupted by people who are in the orchard.

These paths might be wide enough for a lawn tractor but the edge will be hard to maintain, you'll have grasses and flowers growing up to provide a transition from path to forest.

This design will include many "non-productive" plants that are there to support the ecosystem. It would involve leaving some fallen trees to rot in place because this will create habitat and provide food sources for many little critters. The forest islands between the paths may include harvestable crops near the outside edges and have a shelter belt down the middle of it, this area would be like the forest in a "ten" design, providing all of the ecological support without sacrificing the prime edge real estate which you can easily harvest from.

This system cannot be hand-picked but could make for a beautiful You-Pick orchard with walking paths. An "eight" design could also designed as a foraging school or an educational hub.

Canadian Permaculture Legacy on YouTube is a good example of an "eight" design. He has lots of natural areas and his harvest spaces still invite nature into it in a beautiful way.

Seven

Here is where the forest becomes rows. There are no longer large areas of closed-canopy forests in this design, or at least not in the main orchard area.

The orchard might be surrounded by a nature corridor, this is a line of forest that is wide enough to allow the safe passage of animals as large as deer or bears without any real notice from people in the orchard.

This nature corridor also functions as a windbreak for your orchard because this is where we may start switching from full-sized trees down to dwarfed or semi-dwarfed trees. The smaller trees are more susceptible to getting blown over by strong winds because they often have small root systems. This nature corridor serves both us and nature.

The main orchard will have straight rows of trees and would resemble the system that Stefan Sobkowiak uses at Miracle Farms (look up "Stefan Sobkowiak Miracle Farm" on YouTube for reference).

Some of the trees will be fruit trees, some will be nitrogen fixers. There will be bushes and herbs everywhere.

Ideally (in my area) the mulch would be wood chips that are applied once or twice per year. The wood chips would be from smaller branches, up to four inches in diameter, this is better for an orchard than the mulch from a stump or trunk of a tree.

The pathways would be mowed as rarely as possible. This allows wildflowers to bloom, supporting insect life which supports everything further up the food chain. This also creates organic material that can continue to build your soil.

You may consider using methods like sub-soiling between the rows to speed up soil regeneration if the field has been abused for too long. This method creates a cut into the soil, allowing moisture and nutrients to penetrate deeper into the soil, enabling the soil life to come back to life.

You can also add birdhouses to replace some of the habitat that is lost by switching to rows of "productive" plants.

Six

Six will look very similar to Seven but we will be working a bit more towards creating the yields that we want to obtain for ourselves, so we may reduce labour costs by switching to a plastic mulch.

Note: a very durable plastic mulch can last for decades and can have a lower carbon footprint than trucking in wood chips if they are not locally abundant. This means that the plastic mulch could be a part of "Seven". Remember, these are all loose examples.

We may mow a bit more often so the property looks better for eco-tourism opportunities. The mowing could be done in just one part of the orchard though, leaving the rest to function better as an ecosystem.

Some of the flowers around the trees will switch from pollinator support to flowers grown for cut-flower sales.

At this point, we may be using tractors more often, burning more fuel to keep things running quickly.

Five

Five means that we are approaching the point where other decisions are prioritized over the needs of nature. In our example, the business wants to simplify things a bit more and they don't want to deal with quite as many crops.

We remove the herbs and the low-growing fruits like strawberries. But still keep a few bushes per tree, one of these being planted to the southwest of the trunk to reduce southwest injury. We will keep flowers in the mix, ideally a large variety of flowers so there are constant blooms. This supports native pollinators while also maintaining beauty for eco-tourism.

More or all of the alleyways may be mowed. Birdhouses will still be placed frequently throughout the orchard for pest control but with the decrease of herbs, there will be more pest pressure because of the lack of scents to distract or repel them.

Four

Four might involve creating a single row of bushes and flowers in line with the trees instead of having bushes on either side of the trees in addition to the ones between them.

Instead of rows of trees alternating fruit types within the rows, each row may be one species of fruit (apple, pear, etc) but with a few cultivars (Macintosh, Empire, Wolf River, etc.) and still containing nitrogen fixers.

The bushes in each row would produce fruits and there would be various flowers. Still maintaining blooms throughout much of the season to support native pollinators.

At this point, pesticide use will become more common, and occasional fertilization may happen if the nitrogen fixers are too far apart. Pollination services may be required.

Three

We are a long way from foraging in the forest at this point. We have straight rows, with multiple cultivars of apples in one row to limit the spread of pests and disease but are now relying on the application of organic fertilizers and pesticides.

There will still be a single bush to the south-west of each tree, it should flower to support pollinators but may or may not produce fruit. Some flowers could still be included in the space between trees.

We will keep providing habitat for birds in birdhouses and because of the reduced habitat due to the reduction of bushes, we may consider maintaining brush piles for wildlife support.

The near-monoculture of apples could be broken up by adding a few rows of hazelnuts or berry bushes (monoculture rows) to break up the monoculture of apples. By keeping all of the berries in one row it simplifies harvesting, making it a viable second product for the orchard which builds resilience while adding another bloom period for pollinator support.

The orchard will be dependent on pollination services from a beekeeper.

Two

Two is a step up from a standard organic orchard. Instead of having multiple rows of one cultivar of apples grouped together, we will alternate rows of cultivars, One row of Cortland, one row of Gala, etc. We plant the single bush to the southwest of each trunk to reduce time spent painting trunks and to provide a little more pollinator support.

Pathways are maintained just as in a standard organic orchard. We'll still include birdhouses. This is an orchard that is dependent on fertilizers and pesticides in addition to bringing in honeybees for pollination.

One

Here we have a standard organic orchard, it's better than a conventional orchard, not by much, but it is better. Incorporating birdhouses is always a good idea, as is the bush to the Southwest of the trunk.

Summary

Number one is an organic orchard which is a step up from a conventional orchard, we continue to get better until number six when we are establishing a functional ecosystem which is where we start to see some of the best returns because we have reduced our annual inputs like fertilizer and pesticides. While also eliminating pollination expenses.

Our returns drop off around nine and ten because the yields are reduced. This makes the sweet spot somewhere in the six to eight range in which we are improving the ecosystem and improving our personal gains. A true win-win.

I believe that the ideal large-scale commercial design could include an orchard as described in six or seven which also has large areas in the eight to ten range as well.

Large areas of eight could draw crowds as a unique you-pick experience,

Every option on this scale is better than conventional practices, it's just a matter of choosing what works with your values, budget, and skill set.

Once you understand your position on this scale, it is time to set goals.

Chapter Six: Goals

An Amazon search for books about goals returns over 60,000 results. This is a topic that has been written about endlessly. Some of those are by people who are far more qualified to write about goals than I am.

We will make many goals when working on our big projects, but I want you to find the one big goal that will be driven by your values and that will drive the design process that your Permaculture Designer follows.

There are the usual things to say about goals. With some of the famous ones being:

- Make SMART goals
- Use words like "going to" instead of "want to"
- "Make a plan, without a plan, it's a dream"
- Write your goals down
- Keep your goals where you can see them

Some of these have strong merit so I will touch on them briefly and will relate them to some of my own goals as examples.

SMART goals

SMART is an acronym which stands for:

- Specific
- Measurable
- Attainable
- Relevant
- Time-bound

[S]MART

We want a well-defined (specific) goal so we know when we achieved it and so we know what we are working towards and what steps to take to get there. "I want to grow food for people" isn't nearly as specific as "I will grow and give away lots of food to people in need".

With the first option, we may grow more potatoes wheat or carrots, or meat. There are so many directions we can take, but they may not be what you actually want. By saying that I will grow, and give away lots of fruits and nuts I have a clear direction.

This is one of my goals, I now know that I need to plant fruits and nuts in large quantities to make this happen; but how big of quantities?

S[M]ART

We want the goal to be measurable. Saying "I want to give away a lot of food" is great but is a pickup truck load a lot or is a tractor-trailer load a lot? How can I make it more measurable?

By specifying the quantity of 10 tonnes (10,000kg, or 22,000 pounds) I now have a benchmark. I will be weighing the food as it is donated so I can see how close I am to that goal each year as I grow.

I now have a much more specific goal to achieve. "I will grow and give away 10 tonnes of fruits and nuts every year." This is measurable, but can I actually do it?

SM[A]RT

We want our goals to be attainable, if we can't attain them, it's just a pipedream. So, can I grow and give away 10,000kg of fruit and nuts?

Let's do some math. I will likely use primarily semi-dwarf trees because they do well in my area and I don't have to replant them as often. I'm also going to keep it simple and just use one type of fruit now to get a rough idea to see if it's viable.

A semi-dwarf apple tree produces 200-400lbs of fruit per year, we'll call it 300lbs. This means that I would need 74 trees (10,000kg=22,000lbs, 22,000lbs/(300lbs/tree)=74 trees). Using standard spacing of 15' between trees, that would be 1100' of apple tree rows.

It's ideal to have a gap in your rows every 100-150' so we'll just make 150' long rows with 10 trees in each row. We'll put our rows 15' apart. We'll do 8 rows, that's 80 trees in a space that is 150' by 120'. That's less than half an acre. I should be able to increase my goal eventually, but I'll aim for 10 tonnes as a starting goal.

How exciting is it to realize that we can exceed our lofty goals?!?

I plan on a minimum of 6 acres of orchard, I have run the numbers so I know that I can easily afford to give away ½ an acre of production which makes this a financially viable goal! I also have a harvest plan.

Ps. By buying this book you have supported my journey towards this goal as I build my nursery and save up for land!

Will I actually stick to this goal though?

SMA[R]T

If we want to stick to our goals, they need to be relevant, they don't usually explain this part very well "It needs to be relevant to your career" etc, but I think that it needs to be relevant to your values. This is part of why we took the time to explore our values earlier.

So, is my goal relevant to my values? I listed 3 main values, Earth Care, People Care, and Fair Share.

My orchard will be at least 7 on the sustainability scale and will be surrounded by some 8, 9, and 10. This will also be a demonstration site so I can show others how to do what I'm doing.

I am going to restore some abused farmland and turn it into a thriving ecosystem. This is going to be a beautiful transformation that will definitely satisfy my "Earth Care" value, I'll be rewarded as I photograph the birds or listen to them singing as they feed on tent caterpillars in my orchard.

People Care is the main purpose of this goal, I will give away 22,000 pounds of food (I know I keep using different numbers, they are all exciting numbers and I enjoy using all of them to try to understand how much food that is). This is meeting my "People Care" Value.

"Fair Share" is the share of the surplus, in this case, I am growing the main orchard to create a surplus that I can afford to share. I'm creating abundance so I can give away a fair share.

This goal aligns with all three of the values that I mentioned. That's a strong indicator that it's a goal that I should be pursuing and that I will stick to through many trials (like the 300+ trees that were taken out by rabbits and squirrels this past year).

The goal that you have in mind may not align so closely with all of your values, it shouldn't go against any of them though. If you are making a big, life-altering goal, it may be worth ensuring that it does align with all of them.

SMAR[T]
There needs to be a timeline for the goal. "I want to grow and give away 10,000kg of food" is a weak goal. Am I doing it today? Am I doing it this year? Am I giving away that much food over the rest of my life? All of those are valid goals (although "today" and "this year" are not attainable) but the goal isn't specific and I can keep putting it off.

I don't want to give away only 10,000 kg of food in my life; if I live an average life, that's less than 200 kg of food per year, a good goal for a backyard gardener, but it isn't having the impact that I want to have from my orchard.

I want to do this every single year. I won't be able to do it for the first years because I need to buy land, plant my trees, and get them to a productive age.

Buying land is the biggest challenge to this goal, I'm hoping to find land that I can rent until I'm able to purchase it which would let me get started much earlier. If I can find that land in the next year I can plant out my first couple hundred trees as early as next year and have production starting in 5 years after that with it taking a few more years for production to scale up.

I think that I will set an aggressive goal and say that I will do it in 7 years. I'm writing this in 2024 so my goal will be reworded to "I want to grow and give away 10 tonnes of food per year, starting in 2031".

This is a very specific goal, now we will make a quick adjustment to the wording and then make a plan.

You ARE GOING TO Achieve Your Goal

Words are powerful. I used to think this was nonsense but take your goal that you made and change the words between "I want to" and "I will" or "I am going to".

"I WANT to grow and give away 10 tonnes of food per year starting in 2031" is a nice statement in a casual conversation, but when you're doing the hard work of re-planting dozens of trees after the rabbits girdled all of your fruit trees that were almost ready to start producing, or after the neighbour's goats ate your young orchard to the ground. This wording may not be enough to keep you going.

During that challenge I may end up repeating "In seven years I WILL grow and and give away 22,000 pounds of food to people who need it" and it may just give me the motivation to keep going... and to put up that electric fence that I kept putting off.

It's a small difference but it can be a big one. Use the more powerful wording just in case you need it someday when things get tough or when it seems like the finish line is too far away. Your goal can become a mantra in times of need whether it's because something went wrong or because you're picking up an extra shift to pay to take the first big step.

Make a Plan

A goal is a great thing. There's a saying I have heard countless times which is "A goal without a plan is just a dream". I don't agree with this entirely but I do think that it has some merit.

You can have a goal without a plan but you aren't as likely to achieve it. So we want to make a plan. This plan will give you benchmarks along the way so you know that you are making progress and are staying on track.

Some of the parts of the goal may be financial, some may be about education, or about hitting physical milestones like getting the foundation for your new home laid by mid-April so you can move in by Christmas.

Let's continue working through my goal as an example. We now have a well-worded goal that is specific, measurable, attainable, relevant to our values and has a specific timeline.

At this point, we can either choose to work backward from our goal date or work forward from now.

Working Forward

If I owned my big property right now I would work forward. I would start by writing down all of the steps and then would look at how quickly I can get each step done. Each step is like a little goal so make them SMART as well. We'll start just by writing specific steps.

For my goal, I will assume that I have an old farm field that doesn't need to be cleared of trees. My steps would include:

- Create my design
- Prepare the land for planting
- Run irrigation lines
- Plant trees, bushes, and herbs
- Install birdhouses
- Maintain trees
- Make a plan for distributing the food
- Harvest and distribute food

I'm writing this in the winter which is a great time for planning, and it's still early enough to order rootstocks. My budget won't allow me to buy hundreds of mature trees but I can make my own trees for about $10-15 each.

I will create my plan this week, I have already made my observations and learned how to properly design it. I also have decided that I want to plant out one acre per year to keep it attainable, with my overall goal being 6 acres. That's about 200 trees, 600 bushes, 2000 herbs/flowers, and about 100 birdhouses.

> *Step 1: I will create a design for the orchard by Friday. Include plant lists, and material lists for the irrigation system.*
>
> *Step 2: I will order all rootstock, scion wood, grafting materials, cuttings, and seeds for the first acre on Saturday.*
>
> *Step 3: I will order all of the parts for the irrigation system on Saturday.*

I can't do anything to work towards any of the other steps until the ground thaws or seeds come in but I can start making plants as soon as they come in.

> *Step 4: I will graft all of the trees in March and will store them in buckets for Spring planting.*
>
> *Step 5: I will create a schedule and start all seeds at their appropriate times, all indoor planting will be done by May 1st*

With our usual weather here we can't plant out many things until the May long weekend (around the 24th) but I can do a lot of prep work before that point. I need to prep the land and the irrigation system, I can also start building the birdhouses so they are ready to install. Let's add those to the schedule...

You can see how this system works, determining the earliest realistic timeline for a step and writing a goal statement for that step. If I continued with this exercise, I would have my first acre planted by the first week of June and then would start putting up birdhouses throughout the rest of the season.

Note: Planting out freshly grafted trees means I will have to re-plant some next year so I will have to work that into my plan. I could also graft 2-3 years worth of trees and keep them in nursery beds until I am ready to plant out more mature trees.

Working Backwards

I don't own my land yet and I don't have a clear date set for it, so I will work backwards from the end date to establish a deadline for buying my land. Fortunately, a lot of the land around here will work well for this goal so I won't be choosing bad land in a rush to meet this deadline.

Note: When it comes to buying land it is more important to get the right land than to meet your timeline, the wrong land can cause a lot of issues for you.

To be able to give away 10 tonnes of food in 2031 I need 74 trees in full production. I won't reach full production in that timeline so I could also have a few hundred younger trees that are around 5 years old. I will also have many bushes that will be producing much earlier.

A few hundred can be planted in one year and to have 5 year old trees by 2031 means planting them in 2026. I may plant 2 acres in the first year to help get the orchard established quicker.

So The final goal is:

> *"I will grow and give away 10,000KG of food every year starting in 2031"*

In order to achieve that goal I have to set this goal:

> *"I will plant 2 acres of the orchard in 2026"*

In order to achieve that goal I need to set these goals:

> *"I will graft and start 2 acres worth of trees and plants in winter/spring of 2026"*
>
> *AND*
>
> *"I will gain access to my land and prepare it for planting in 2026"*

I now know that I have a two-year timeline for finding land, I don't need to have purchased it in that time but I need to have secured a deal and be able to start planting it out even if I'm renting it in the meantime with a contract to buy it in the future.

I can also get a bit of a head start by growing things in my nursery for a year or two before planting them. So, while I don't need it to meet my goal, I will add this step/goal just because I can:

> *"I will order rootstocks, scion wood, cuttings, and seeds on Saturday to build my nursery so I can plant 2 acres of trees in 2026"*

We can work forward or backward, both are valid options depending on your preferences and situation. Either way, we must make a plan. If you don't have all of the information, you can leave some gaps and adjust the plan once you know what you need to know.

For example, if you want to build a permaculture property with goats for soap making but you won't be buying land until next year, your plan could look like this:

Step 1: I will create a plan today to save money to purchase land next year.

Step 2: I will work through this book to clarify my values, set goals, and create a plan by next Saturday.

*Step 3: By the end of the month I will sign
up for the next available soap-making classes from
two different businesses.*

*Step 4: This Summer I will find a goat farmer with
similar values to learn from and to mentor me.*

Step 5: I will buy land by March of next year.

*Step 6: Within 1 month of purchasing my land
I will reach out to my Permaculture Designer
to begin the process of planning our property.*

*Step 7: Create a new plan once I have met with
the Permaculture Designer.*

Steps 1 through 4 are preparing you for your goal, allowing you to build skills, knowledge, resources, and relationships to prepare you for when you have your land.

Step 5 is what may have been step 1 for most people's plans but it makes sense to, as Jess Sowards (from Roots and Refuge) says "Turn your waiting room into a classroom". (Sowards, 2021)

Step 6 sets a timeline for beginning your design work. It can be best to wait a year so you get to know your land but your designer may be able to help you create a plan in stages that can be modified if you find that the pasture is flooded for the entire spring, or that there are other factors that require a change in the plan.

Step 7 seems a little obvious, but it is good to write it down with a timeline because it will make sure that you don't proceed without proper planning. Once you have talked to your designer you will be able to start looking into buying your goats and your fencing; planning the steps for putting the design into practice.

Make your plan

It's time to make your plan, but only if you have completed the inner work of defining your values, and deciding how regenerative you want to be, and also have defined your goals.

Write Your Goals Where You Can See Them

I'm combining two here for the sake of being brief. We are making these goals, we are wording them well, we even know how we will achieve them.

Now write your goal down and put it somewhere that you will see it. This could be the background of your phone or computer, or it could be on your fridge, I had mine on the inside of the visor in my car.

Seeing them helps keep them top of mind. I find that I start to ignore things that I see every day, so if you are like me, you may do well with having them in places like the visor of the car or the inside of a closet door so you don't get used to seeing it. I notice my goals when I flip my visor and it makes me think of them more frequently. I don't flip my visor often so I notice the paper when I do, I also intentionally look at my goals regularly.

You may work differently than me so find what works for you and do that. As far as what goes on the paper, you can use text by itself or you can include a picture or even a collage of photos. Whatever you do, make sure you include your well-worded goal and that you remember why you set that goal.

You are now on the path towards achieving your goals instead of having some vague longing. The rest of this book will be about identifying your resources and communicating with your designer then implementing the design after it has been completed.

Chapter Seven: Eight Forms of Capital

When we think about raising capital we tend to think of raising money, that is the only form of capital that most North Americans are aware of.

Ethan Roland from Appleseed Permaculture is the first one I can find who described the 8 forms of capital as they have been accepted in the Permaculture world. These are the forms of capital that create resilient wealth and they factor into how you will proceed with your project.

<u>The eight forms of capital are:</u>

Financial Capital
Material Capital
Social Capital
Spiritual Capital
Natural Capital
Time as Capital
Intellectual Capital
Experiential Capital

(Roland & Landua, 2015)

Time

In most illustrations, we place time at the center because without it we can do nothing. Time has 2 primary meanings. One is the time you have left in your life, and the other is your day-to-day free time.

If you are starting a project at 90 years old you may have a lot of time available in your day but likely don't have a lot of time left overall to work on your project. On the other hand, if you are starting at 25 but are working two jobs to get by, you may (or may not) have a lot of time left to live but have very little time to invest in your project.

Understanding your time and being realistic about it when creating your plan can be the difference between a successful plan and a total flop (which would give you experiential capital but that's not what we are looking for).

If you are young and don't have much financial capital, you can look at growing your trees from seed and through grafting because you can afford the extra year or two. If you need your trees to start producing sooner you will likely end up exchanging other forms of capital for mature trees.

Your resources of time can also dictate whether your project can be built in slow phases or whether a lot of it needs to be installed all at once.

 This limitation on your time can be related to external factors as well. If you are building a farm on your property but you foresee the nearby city expanding towards you, there is an opportunity to have an excellent market but there may also be limitations that pop up as the area around you changes from farmland to a new subdivision. All of the sudden you may no longer be able to apply for agricultural zoning or may not be able to build the outbuildings or start raising livestock. These may be reasons to start certain projects earlier than you would otherwise do because your time is limited for those projects.

Some people who approach me are looking for a second set of eyes to walk through their property so they can get a confirmation that they are on the right path and are making the best use of the space. The majority of these people plan on using their own time to install projects and work the land

Some people who approach me want a one-stop shop. They want the plan to be created and installed for them because they don't have the time, or don't want to spend their time on the installation. They may also have physical limitations with their health that require them to hire the work.

Your time is also something that you can trade. You can trade your time for money, materials, and other forms of capital. One valuable way to exchange time is to exchange it for Intellectual and Experiential capital through internships, workshops, or employment /volunteering for someone who is already doing what you want to do.

In some lists of the eight forms of capital this is replaced with Cultural Capital. Our songs, stories, traditions, and history are cultural capital. These are passed down through the generations and have much to offer.

Financial Capital

Financial capital is something that I am familiar with, I have a background in finance and have seen how poor our understanding of and relationships with money is in our culture.

I don't like how much we prioritize financial capital, yet it is the longest part of this chapter because there are key lessons that everyone can benefit from, especially if you are setting out on a new project.

Our parents generally didn't feel confident in their knowledge of finances, so they are hesitant to teach us. Many of them hoped that we would either learn about it in school or would do a better job of figuring it out than they did. Unfortunately, I have seen that this isn't the norm; most school systems aren't teaching us about finance. I reached out to many of our local high schools and none of them had any interest in having me teach basic finance to their students, even as a free service.

I say this to let you know that you are not alone if you haven't learned to budget, invest, or pay off debt. These are important skills to learn so I encourage you to learn them so you can make the most of the financial capital that you do earn. I've been intending to re-film the finance course that I designed while working in the finance sector but do it with a homesteading focus. If you are interested in this course reach out and bug me about it so I make the time for it!

Our relationships with money are also a little messed up here. We live in a culture that doesn't find the balance between saving, spending, and giving (the three uses for money).

Savers
Some people save everything at the expense of the other two uses for money. They want to have their millions of dollars and don't worry about what it costs. This can lead to underpaid employees who are working two jobs just to get by or to saver experiencing a lack of joy in their life. The benefits of saving are that we can have a greater sense of financial security and that we have the money to be able to make our big visions come to fruition.

Spenders
Some people spend everything with no consideration for saving or giving. This usually results in buying massive quantities of low-quality products that may have been manufactured with slave labour. This leaves nothing for a rainy day or for when we see others in need. The benefit of this is stimulating the economy. Spending also pays our bills like our mortgages, our taxes, and our food bills so we have the things that we need.

Givers
Some people give everything away, leaving nothing to spend or save. These people often end up in some of the hardest places financially. If you are a giver you may find yourself thinking "They need it more than I do" and that's okay, so long as you are being honest and they actually do need it more than you do; this means that you have paid for your food, shelter, utilities, transportation, and have a plan to make sure you can continue to pay for these things as you transition into retirement.

The practice of giving
Giving is the one that most people struggle with and is one of the reasons that I like the idea in the Church of tithing. This is the practice of giving a tenth of what you earn to the Church so they can do their work of caring for the wider community. I know that this is not the experience that many have of churches, and we hear the stories of pastors of mega-churches with private jets. It is important to give money to where it can be used well.

If you are not religious, I would suggest practicing tithing, but giving money to charities or keeping an envelope in your purse or wallet that you can give to someone in need instead of to a church. This can transform our relationships with money and can be especially good for the aggressive saver. I have seen savers become less stressed about money when they view the world through a lens of generosity. It can also help the spender as they see a better use for their money than buying yet another toy.

In an ideal world, we will balance these three uses of money. Your plan to achieve your goals would be served well by considering all three uses for your financial capital.

Assessing Your Financial Capital

How much money do you have right now? Did you only think of the cash? Did your mind calculate how much room you have left on your credit card? On your available overdraft? In your line of credit? Did you think of the money that you have been saving for your new home, boat, or for retirement? What about the value of your possessions?

How much money will you have? Do you have an inheritance or settlement coming in? Are you going to sell your house to start your new project? Is your income going to change in the near future? Will you be making money from your property? If so, how much? What can you sell or do to get some cash?

What is taking up your money? Do you have a mortgage? Car loans? Child support payments? Credit cards? Payments on your laptop? Subscriptions for streaming services, internet service, software, etc?

Cash Flow

Understanding how much money is coming in and how much is leaving is the usual job for a budget and I think that's a great starting point for a budget. Start by just writing how much money is coming in and out, be honest, and do your best to do it well.

The real job of a budget is to tell your money what to do. Once you understand what your money is doing right now you can compare that to your values and goals and start telling it to do something else. This is an essential tool for getting the most out of your money. I don't really care about money which is why I want to make sure it does what it needs to do efficiently so I don't have to earn more than is necessary.

Your cash flow will affect how quickly you can achieve your goals. Either it will define how much cash you have available for spending, or it can affect how much debt you can access and pay off.

Debt

I'm a fan of avoiding debt, but not everyone has the same priority or value around finances so I will write about it.

We have access to many forms of debt. We can use credit cards, can get payment plans for just about anything, can have car loans, and mortgages, and can borrow money against our homes.

The ability to access debt is limited or enabled by a variety of factors. What you are using the money for, how much money you earn, how much debt you already have access to (even if you aren't using it), your history with debt, your employment history, and more.

Debt can be a tool that some people like to use. I used debt as a tool to buy my first house. That house is a duplex that is helping fund my retirement; I will pay off that debt as quickly as possible.

I know exactly how much money I owe and it weighs on my mind because that debt is a risk factor in my plans. If something happens to my income, I could lose my home because I owe money on it. I value the security of knowing that my house is safe, this is why I will pay it off.

Make sure you factor debt payments into your future cash flow. Avoiding debt is the lowest-stress option, while planning for debt alleviates some of the stress that comes with carrying debt. Spending every dollar you can access to build your project is of little use if you have to take another job just to keep the doors open and end up burnt out.

Savings

We can save money for many goals and purposes. The most common things that we save for are:

Buying a house

Buying a vehicle

Retirement

Vacation

Travel

Emergencies

Renovations

Repairs

Before using your savings for a project, consider the intended use for that money as well as the benefits and consequences of using it.

An example is saving up for a new roof. You probably shouldn't use that money to pay your designer or to buy a herd of goats for your soap business. On the other hand, if you are saving for a boat or vacation and are willing to sacrifice that luxury, you can use those savings however you like.

Sinking Funds

Sinking funds are a powerful tool that I will touch on very briefly.

A sinking fund is an account that you create and add money to for a specific purpose. Use accounts that earn interest and have no account fees for your sinking funds. For long-term sinking funds you may be able to invest the money, that comes with other risks and is beyond the scope of this book.

I have a sinking fund for my new roof because it is a major purchase that I will need to make in the next 5-7 years.

How it works:
Your roof will cost about $10,000 in 5 years
$10,000/5 years is $2000 per year that you need to save.
Monthly contributions are easier so $2000/12 months is $166.67/month

In 5 years you'll have all the money that you need to pay for that roof with cash. You may check the price again in 2 or 3 years to make sure you are still on track and that you estimated the inflation properly.

Making payments before the purchase allows us the flexibility of skipping a payment if there is an emergency, an option that we don't typically have when paying off a debt.

Consider applications for sinking funds. Paying your designer or paying for a new barn or for a professional marketing company to rebrand your business are all options within the realm that this book covers. Saving up for a camera was my first sinking fund as an 11-year-old, whereas saving up for a used car was my first sinking fund as an adult.

Emergency Funds

Everyone should have an emergency fund that is for emergencies only (not to be dipped into for a vacation or Christmas). This is especially important for anyone who owns a home, farm, or business.

Some people have an emergency credit card, which is an option but I'm not fond of it because they can get cancelled right when they were needed most. In 2008 most lenders in North America canceled or reduced limits on credit cards right as everyone was in a pinch... Because everyone was in a pinch.

Credit cards can also get cancelled or have their limits reduced if you lose your primary source of income which is one time when you may be most dependent on an emergency fund.

I like Dave Ramsey's approach to emergency funds. Step 1 in his 7-step process is to create a small $1000 emergency fund, this isn't enough to replace your roof but is enough to buy a tarp or pay your insurance deductible (have a $1500 initial emergency fund if you have a $1500 deductible). This can also cover small emergencies like getting two flats at once. (Ramsey, 2003)

Step 2 is to pay off your debts other than the mortgage, then step 3 is to create a 3–6-month emergency fund. This isn't 3-6 months of income, it's 3-6 months of expenses. This enables you a six-month buffer to heal from sickness/injury, to grieve the loss of a child, or to gain skills required to find new employment.

As a homeowner, this allows for the replacement of your hot water tank and furnace the same week that your teenager crashed the car and you had to use your insurance. You could do all of this without panicking and may still have money in your fund.

As a farmer, this could cover an emergency vet bill or to replace your fences after your cattle get spooked and charge through three paddocks without slowing down for fences.

Other funds should cover some of these costs. For example, a maintenance fund covers the replacement cost of a hot water tank. The emergency fund is our last line of defense for when the tank fails earlier than expected.

A good emergency fund allows us to take bigger risks too. If we are always one emergency away from losing everything, we may not have the emotional capacity to start a new project, even if the risks involved are very small.

Emergency funds should always be "liquid". This is a term that means readily available. Cash is liquid, and so is money in a savings account. Gold isn't liquid, most investments aren't liquid, and business inventory likely isn't liquid enough for this purpose.

Many people want to put their emergency fund into investments so their money can work for them, that's fine for retirement savings, but the work that your emergency fund is doing is to protect you from having to put the new furnace or veterinarian bill on a credit card that charges 20% interest.

Consider building and maintaining an emergency fund.

Future Money

Sometimes we know that a large sum of money is coming and can plan for it. If you know that you will be selling your house, you can choose to factor that money into your project during a stage of the project that makes sense with the timing of the sale of the house.

Getting a design made while knowing that you will likely receive an inheritance in the future enables you to include elements in the plan that will be built at a later stage when the money arrives. These elements can be planned for now, so everything works smoothly. An example of this would be if a parent is likely to pass soon and leave you a large enheritance.

While we may not want our parents to die, we can still think ahead to what that will look like. Your parents may even be excited to learn how you will use the money that they leave behind. Will you buy the field next door and expand your project? Will you build a few cabins to start a retreat center? If my parents were leaving a large Inheritance, they would love to see what will come of it and to know that it wouldn't be wasted on bad decisions.

You may have other money coming, your design may be for a property that has a black walnut timber farm on it and you have a buyer lined up for some of the logs. They might harvest the trees and pay you for the logs once you take possession of the property. This plan to transition natural capital to financial capital can be factored into your overall plan.

Income from the results of your project can also be considered but we need to be realistic and to make sure that you aren't so dependent on it that a delay puts your project at risk.

When I bought my house I took out a short-term loan. I ended up running into issues with the building department. They wouldn't respond to any of my phone calls, emails, or letters as I tried to get a permit to start construction. It took six months to get my building permit, which was supposed to be when I'd be switching to a traditional mortgage with a finished house. I didn't know that my township would take 1,300 percent of the maximum allowable time to get me a permit and I didn't have a plan that would allow for this delay. My saving grace was a gracious lender who was also going through building department issues on one of their projects.

Even with that kindness, it took me years to recover from the town's delay. I had additional expenses including heating an uninsulated 3000-square-foot building in Canada and had decreased income because I couldn't rent out the unit when intended. A large portion of my renovation budget was eaten up during that time. Learn from my Experiential Capital so you don't make the same mistakes.

Material Capital

Material Capital includes everything from buildings to piles of metal. What do you have? Do you have a pile of scrap metal? Do you have an old barn? Do you have the roofing metal from an old barn? Or timbers from that barn? Do you have piles of plumbing materials? Is there a pile of stones that you think of as a tripping hazard? These may be able to be used or transformed into another form of capital that is useful.

This form of capital also includes your tools. You may list in your intellectual and experiential capital that you know how to operate a sawmill for creating lumber and timbers, but do you have one or have access to the equipment?

Social Capital

Social capital is goodwill, it is favours that you owe, and that you are owed, these are connections that you have. Do you know the building inspector personally and will be able to make sure you get your permit right away? That's a bit of social capital that I would have benefited from.

After six months of trying I used the very small social capital I had with the mayor (just being friends on Facebook) to look for a solution... It resulted in a complete turnover of staff in that department. That (now ex) mayor has some social capital with me that he can collect on and that I will gladly pay!

This isn't about leveraging people or using people. I am aware of social capital and use it frequently within the constraints of my ethic of People Care and loving people well. This is about community.

Who do you have in your community? Do you have a friend with a tractor for that one-time plowing of your field? Are you a member of an Earthship building group that is filled with people who want to experience a build and who you can call upon them to help you pack tires as you build your home?

I was talking to a friend about hiking to the North Pole (before we saw the cost), and he said "Maybe your book will sell 1,000,000 copies and you can pay for both of us to go". He has supported me well for most of my life and has built up enough social capital that I'd probably do it.

He also said that he would write a book and see who sells more. I started jokingly naming people I would send a copy of my book to so they could help me with marketing. As I was creating the list, I realized that I would not have considered this social capital at all if I wasn't jokingly name-dropping people I have been lucky enough to meet through my permaculture journey.

It turns out that I know some very well-connected people. I hadn't considered it at all, I hadn't thought of them as resources other than as teachers or colleagues. I expect that they would all be excited to support me and that it wouldn't use up any social capital but would instead increase the social capital that we have together as I share my journey with them and as they share their feedback with me. Maybe they will even think that it is good enough to pass along to someone else.

Make a list. List some of the people you have massive amounts of social capital with like my buddy Jon who I'd take to the North Pole just to spend time with. Or another friend who, at one point, I considered moving closer to and offering him free rent just so I could be in his presence more often.

List some people with whom you have some social connection like your teacher from high school, a coworker, or the photographer who shot your wedding 9 years ago.

Think about the people you have the tiniest connection with, like that mayor that I had on Facebook. I don't remember when we added each other but that was social capital that I was able to leverage for advice; he did much more than give advice and I'm grateful for that. Maybe you get deliveries at work from someone with a truck; or you wave to the guy who clears the snow from your work parking lot with his big Cat loader, maybe you bought him coffee a few times.

We aren't going to use these people, that is a way to burn bridges and be a generally miserable human being. We may reach out to them and ask for a hand or we may have something to offer them.

Spiritual Capital

I find Spiritual Capital to be hard to define and a search on the internet and through my bookshelf shows that I'm not alone.

Spiritual capital has been defined as our ability to be our authentic self, as well as being our spiritual knowledge/expertise; and as the capacities that we build through spirituality.

I'm still learning, but I feel as though it is more important than we give it credit for. Spiritual capital is what I relied on when I was alone on the streets in Australia with a paralyzed arm, no money, and no hope aside from prayer. In that moment, my faith (spiritual capital) was far more valuable to me than money could have been. I ended up using the last of my financial capital to buy more spiritual capital in the form of a bible. That decision is one of the best that I have ever made.

Natural Capital

Natural capital is all of the living things around us and the effects of the ecosystem as a whole. This includes plants, trees, soil, and purification of air and water. Plus the songbirds, cattle, even the birds of prey that come for your chickens, and the rats that eat their food. These are all forms of natural capital.

Intellectual Capital

What we know, what we have learned, and all the information that we have retained is Intellectual Capital. What do you remember from school? What did your parents teach you? Have you taken courses?

I have invested heavily in intellectual capital, I use it when doing design work. I use it when fixing or building things, I use it when teaching workshops or public speaking, and I also use it as I write this book.

Experiential Capital

Experiential Capital is your street smarts. What have you learned through your mistakes? What have you learned through your successes? You may have studied welding, sewing or baking, but what was that one experience that taught you to create a better product?

I like to compare it to learning to bake bread with my grandma, the recipe was important but what sets her bread apart is the little secrets like adding flour "until the dough feels just like this". That experience of knowing how the dough should feel is experiential Capital.

Chapter Eight: Opportunities

We looked at the eight forms of capital (Financial, Material, Social, Spiritual, Natural, Intellectual, Experiential, and Time). In this chapter, we will be looking at those in a different way to identify potential opportunities.

Some of the opportunities are simple, you can exchange financial capital for material capital. We do this when we go to the store to buy groceries or a shovel.

Sometimes it is simple but isn't as plain to see. When you pay your Permaculture Designer to design your property you may think that you are trading financial capital for material capital, which is the physical design. While, in reality, when you purchase a design you are trading Financial Capital for multiple things, the ones that come to mind are:

-Material Capital (physical documents)
-Intellectual Capital (knowledge and training)
-Experiential capital (lessons learned from previous design projects and experiments)

-Social Capital (I want you to succeed)
-Natural capital (if I am providing plants)

We can trade more than money. As you can see, I make my living by trading things other than money for money. You likely do the same when you work.

We also make exchanges while excluding financial capital from the entire conversation. This can be bartering but this is also spending seeds (Natural Capital) and time (Time) to increase our natural capital in the form of plants, or, if that seed was a black locust, cedar, maple, walnut, oak, etc. we can turn it into material capital in the form of lumber which can be exchanged for many forms of capital or transformed into infrastructure like a barn or chicken coop (more material capital).

You may not need to start with the seed if you do a thorough assessment of the opportunities that you have. You may have seedlings growing on the edge of the field that you can nurture. Let's look through some opportunities in the various forms of capital that we talked about earlier.

Time

The opportunities in time are as multitudinous as the ways that we define time. Below I show situations involving time and the opportunities that they may bring.

Free time	Time to learn, to build slowly, to do work yourself to save money, time to trade for money, etc
Old age	Pressure to avoid wasting time and cut out what isn't important to your plan.
Young age	Time to make mistakes and learn from them, time to grow something really big, time to slow down.
Busy life	Motivation to seek out efficiencies

Time is a limited resource, yet it provides opportunities in every situation. The busy person may find the method that saves countless hours of work. The person with all the time in the world may be able to take the time to build their house themself to shave years of waiting off of their schedule because they don't need to pay for labour.

Look at your time situations and see how you can leverage them. Are you stuck waiting to pick your kid up from school for two hours every weekday?

Spend that time reading, resting, learning, and doing anything other than mindlessly scrolling.

Are you stuck working alone at a mindless job for eight hours per day and are allowed to listen to music? Consider listening to audiobooks instead. Many libraries have a database of audiobooks that you can listen to on your phone. I listen to an average of 10.5 hours of audiobooks per day, they are a mixture of ones that teach me something and ones that let me relax my mind as they take me on an adventure.

Seek your opportunities, especially in the problems. One of my favourite permaculture phrases is "The problem is the solution". In the last example about working alone, you may think that the time you spend at work is a problem but it may be the solution for you not being able to find the time to gain intellectual capital.

A classic example of "the problem is the solution" is

"*you don't have a snail problem, you have a duck deficiency*"

– Bill Mollison (Mollison)

The pesty snails that were eating your garden are food for the ducks. Find the solutions in your problems.

Financial Capital

You have already found one opportunity from your financial capital, you exchanged it for this book and for the intellectual capital that it will give you. Here are some other examples:

Money but no time	Hire help
Money but no materials	Buy materials
Excess money	Invest in people, business, or to grow your financial capital
Money deficiency	Sell things and simplify life
Money deficiency	Pressure to get creative or to learn skills
Money but no skills	Pay for courses, and workshops, or have others complete projects or designs for you
Money but no desire	Pay someone else to do the undesirable work

Again, you can find opportunities in both an abundance of and in a lack of this form of capital. I learned an astonishing amount of skills because of a lack of capital. For example, I wouldn't have learned all of the skills that I needed in order to start my nursery if I had the financial capital to exchange for all of the plants that I wanted.

Material Capital

Got Junk? Use it! We live in a world that is covered in material capital. Did you know that 92,000,000 tonnes of fabrics go to landfill every year? If you have a clothing factory nearby that produces cotton clothing, how can you collect and use that waste? Cotton mulch? Insulation? Biochar? The opportunity here is to get resources for free or for very little money and give them a second life while saving them from landfills. Fast fashion is estimated to cause up to ten percent of global emissions, and a lot of this happens after disposal, you would be preventing harm to the environment while getting resources!

Here are some examples of Material Capital and how it can be used:

Empty barn	Contain livestock, store vehicles, convert to housing for workers, WOOFERs, or rental income
Old rolls of fencing	Build a Bioreactor compost, or potato tower
Used cooking oil	Biofuels for diesel tractor
Old trailer	Chicken tractor
The farm came with old furniture	Clean and sell it
An old solar panel that isn't efficient	Connect it directly to a 12V fan to automatically vent your greenhouse when it is sunny
Pile of lumber	Trade for an abandoned greenhouse frame
Pine trees	Logs for a log cabin, or can be milled for timber frames

Whether using it on your property, or trading it for something you can use; look at every bit of waste and see if there is a better use before you trip over it yet another time or send it to a landfill.

I have traded lumber from the sawmill for resources of many types. I also have two old solar panels that still put out enough juice to run a small pump, a bubbler, or a fan. They have been carefully tucked away to be used this spring. Find the opportunities that you have lying around.

Social Capital

Community is a beautiful thing when we engage in it properly. I didn't used to engage in it properly and it limited the benefits that I experienced, including the depth of connection and the support that I would have received from my community.

At every chance, I would help others. Whether a friend needed someone to sit with them through a tough season of life, or needed help moving or renovating their house, I was there. I didn't let others do the same thing for me though. If I was going through a hard time I felt that I had to get through it on my own.

When I spent years renovating my house, I only invited one friend over to help one day. I had family who helped (especially my amazing mother who just told me when she would be there and showed up). I didn't ask my friends to help. Looking back, asking them to help would have resulted in good memories, deeper connections, and a deepening of community.

Because of this, I'm working on using social capital as a way of investing in building a sense of community. By allowing friends to help me put up a greenhouse we would be spending time together as they experience something new. Later, we would light a fire and sing songs while enjoying food from the garden (fire-roasted veggie skewers anyone?).

By inviting friends to help me I am also leading the way for them to invite me to help them.

I write about this for those who, like me, don't like receiving from other people, and who don't want to burden others. This is an essential part of community. I believe it was Brene Brown who spoke on power dynamics in relationships. If I am always helping someone else, but never give them a chance to help me, it creates a strange dynamic where I have made the other person feel like a burden, incapable of contributing to the relationship.

With that knowledge, let's look at ways that we can leverage this social capital.

Your friend is a beekeeper and you want to learn	Offer to meet them in the bee yard regularly to help. For the first few days, you may slow them down but we beekeepers generally like to share about beekeeping. After a few inspections, you may be able to start inspecting a few colonies on your own, calling your mentor over when you have questions. This relationship becomes more mutually beneficial the longer it lasts.
Your neighbour has a tractor and you need a field tilled before converting it to an orchard.	Ask your neighbour if they can help. Recognize that their time, their vehicle's wear, and fuel are all valuable. I know some farmers who would love to spend an extra hour in the tractor helping someone out and would do it for a loaf of fresh bread. I know others who are just getting by and would do it at a rate that earns them money while costing you less than renting a tractor and implements.
You are campaigning to convert a vacant municipal lot into a community garden.	Reach out to Council members who represent you, to business owners who may donate materials, and to community members to campaign with you. Reach out to others who have completed similar projects, the fact that you are passionate about similar projects can generate instant social capital.
You need help raising a barn.	It's time for a BBQ and to send an invite. Make a day of it, maybe even turn it into a potluck. Your friends will talk about that time they built a barn in one day, for years!
You need a cup of sugar for a recipe.	Knock on your neighbour's door.

The Sugar interaction is so classic that we use it to define good neighbourhoods, yet it is just one person asking another for help. It is my favourite example of social capital because it is so simple.

I remember reading many variations on a story that may be fiction but that we can learn a good lesson from. A child asked her mom something along the lines of "Why are you asking Miss Potts for a teaspoon of salt when we have lots?", and the mom responded "Miss Potts doesn't have much, but I know that she has salt. She has to ask us for help sometimes so I like to ask her for something small from time to time so she feels like she gets to help out too. It's our little secret, okay?".

The mother in this story knew that it isn't healthy for anyone involved if one person is always the one giving in a relationship. It's not about the salt, it's about allowing Miss Potts to contribute to the relationship. Miss Potts doesn't need to know that the mother has lots of salt, she just needs to know that the mother needed salt for a recipe and turned to her for help.

It's best if the need is real, even if it's small, but I like this story and the message that it spreads.

Now, if you are inviting people over to help you do something like putting up a greenhouse and having a bonfire, don't forget to mention the greenhouse part. Don't bait people in with promises of smores then make them work for it, this will have the opposite effect of asking Miss Potts for salt. This will breach their trust; this is a mistake that is common in the Multi-Level-Marketing world.

There are people who I won't get together with because I know that it is too likely to be a sales meeting, instead of two friends getting a hot drink on a winter day. If they actually needed me now, they would have to convince me to be there for them, not because I don't want to help but because my guard is up now, I'm always expecting the sales pitch.

Use Social Capital well, it will make your life far more beautiful than it ever could have been without using it.

Spiritual Capital

Not everyone is spiritual but I believe that we are all spiritual beings having a physical experience.

As mentioned in the chapter on the Eight forms of capital, I struggle to define and explain this form of capital. I will instead write about some examples of how I would seek to use and request spiritual capital.

If I'm facing a major decision or a tough challenge, I will immediately pull on my social capital with my Grandma. She can use some of her Spiritual Capital as she prays for me. I know that she prays for me regularly but by giving a specific request she can lift that need up in her prayers. She may even be talking to someone and realize that connecting us would be the solution for each of our respective problems.

When I was living on the streets in Australia with a paralyzed arm, I invested in spiritual capital when I dragged myself and nearly 100 pounds of gear across the city to buy a bible and to read it to center myself and remember what is important in life.

I think that the processes we have walked through in this book build spiritual capital.

> By defining your values you define what is important to you through every season, this gives you an anchor.

> By defining your goals you ensure that you have purpose.

> By defining your plan and writing it down you give yourself confidence in your ability to complete this task.

Examine your life for spiritual capital and please reach out if you have a revelation. I'd love to learn more about this form of capital.

Natural Capital

Have you ever seen the return on investment when investing other forms of capital into natural capital? Nature is amazing.

I received about 100 rhubarb seeds for $8 and planted them all in the middle of winter, it was a treat to have all of that green to enjoy inside my apartment. That spring I planted out the 86 plants that germinated and survived my inattentiveness. The following year I allowed a few of the nicest plants to go to seed. I collected approximately 24,000 seeds. That is approximately a 1,500 percent return on investment per year. If I continued to plant all of the seeds for five years (from the time I plated the first 100) I would end up with about 100 million rhubarb plants. This is remembering that I only allowed about 8 percent of the plants to produce seed.

For comparison, the stock market averages about 10% per year in the long run. If we invested $100 instead of 100 seeds we would have $161 at the end of that same five years.

Nature is a powerful force and I enjoy working with it. It provides more than just plants. Here are some examples of how we can leverage Natural Capital.

The sun shines on your property	Convert it into electricity using a solar array. Or convert it to heat for your house using intentional design.
Grape vines that need pruning	Cut the pruned material into 6-12" lengths and push them into the ground, add water, sunlight, and time as they turn into new vines that you can plant out.
Noise from a nearby highway	Plant carefully arranged trees and bushes to reduce the sound to a barely perceptible level
Oyster Mushrooms	Add to an oil spill to convert the oil into material that supports life again.
The soil under your greenhouse	Insulate around it and run pipes through it. Blow the hot summer sun through these pipes to cool the air by transferring heat to the soil. In the winter you can turn on the same fans to pull the heat from the ground back into the greenhouse to heat it.
Cattails on the side of the road.	Add them to your pond to help filter the water while producing habitat and food sources.
Snails eating your plants	Add ducks that will feed on the snails and turn them into eggs,

	meat, or more ducks.
A pile of rocks	Use for landscaping, stone wall construction, etc.
Trees where you want a garden	Use the wood for fences or raised beds. Chip the branches for mulch. If the wood is high value you may be able to sell it to cover the costs of stump removal and other projects. Make a pole barn. Firewood.
A South-facing cliff	Creates a microclimate that warms earlier in the Spring and stays warm longer into the Autumn.

Natural Capital often presents itself in the form of problems like a pile of rocks that you can't mow around. Those problems can be turned into amazing solutions.

My parents no longer want to grow raspberries so they have a large patch to remove. This patch is a thorny mess that is preventing them from replacing their shed. That patch contains hundreds of dollars in the form of raspberry plants.

I will be planting them out in my berry patch so I can continue to grow them out with the intention of relocating them to the farm someday. They would use their Social Capital to get some help removing them either way. In this scenario, I will accept a payment of Natural Capital in exchange for my time. The Social Capital is being exchanged for more Social Capital.

What opportunities can you find in your Natural Capital?

Intellectual Capital

Francis Bacon famously stated that knowledge is power (Bacon, 1597). How can we find opportunities in our knowledge?

I use my knowledge of Permaculture Design to create designs that are changing the world one plot of land at a time. My 10th grade English teacher Mr. Kingsbury used his knowledge to create a career that he then used to invest in his students. He also incorporated experiential and social capital as he invested in us. I don't think that he ever would have expected me to write a book but he taught me some of the more valuable lessons that I learned in high school and they had nothing to do with English. If it wasn't for his Intellectual Capital in the subjects of English and Education, he wouldn't have had the opportunity to shape my life.

Come to think of it, I may share some of my intellectual capital (a copy of this book) with him as a demonstration of the social capital that he has in me.

How can you find opportunities in intellectual capital?

Gardening knowledge	Plan your garden and grow food for yourself and/or for others
Mechanical knowledge	Fix or build things for your property to make it run more efficiently. Repair vehicles from home for extra cash.
Seed starting knowledge	Start your own seeds, sell seedlings, or teach a seed-starting workshop.
Engineering knowledge	Design your own projects to save money. Work with your Permaculture designer to mesh their intellectual capital with yours to create more detailed plans.
Data processing knowledge	Processing data can allow you to work remotely, even while travelling or from your new greenhouse. Data is also an excellent tool for understanding operations on your property; it's not essential but you can learn a lot from it!
Video production knowledge	Create educational or entertaining videos (I just checked and there are "Chicken TV" videos with 10,000-2,400,000 views, it's just videos of chickens). Create marketing content as a side gig to allow you to transition away from full-time work earlier.
Social services and youth-work	Build systems into your project that allow youth to learn and explore while having a safe place to make

	mistakes and experience failure. A place where they can find a new path for their life.

Your intellectual capital may not seem like much to you, but if you can learn how to turn vegetable oil into fuel for your tractor you can save money or can use it as part of your marketing to display how eco-friendly your property is.

We often overlook our own strengths, this may be another good time to use some social capital and ask others who know you well.

Another note on intellectual capital is that we can learn a lot while teaching others. We often learn even more through teaching the topic to children.

If you know more than most people, it may be your time to start teaching. You can be honest about your knowledge level, don't say that you are an expert master gardener if you are a backyard gardener. Don't downplay how much more you know than the people who think that strawberries grow on trees (this example may be a jab at my very intelligent brother who made this mistake as a kid, he'll never live it down).

Share your intellectual capital and use it in many other ways.

Experiential Capital

Experience is one of life's greatest teachers, some of the knowledge that it teaches us ends up as intellectual capital, yet some of it is different. How do you explain how you know that the pathway should be curved a little differently? How do you explain how you know that you need an extra third cup of water in your bread because it feels like a dry day today?

This is experiential capital and there is only one way to get it, (yes, I'm going to say it), through experience!

This experience can come in multiple forms. We can make mistakes, we can have successes, and we can have someone show us the difference in sound between a calm colony of bees and an angry one (a lesson that we may not forget for a long time!). We can also learn something because someone showed us; grandma taught me to knead this dough until it feels "just like this".

Finding the opportunities to list in this category is tricky because it can be about the nuance that makes the difference between something good and something great.

The good chef may know that they need to cook their steak for a certain amount of time on each side to deliver a medium rare steak to their customer. This result may be pretty consistent but there are some variables that affect how quickly the steak cooks.

An excellent chef may know that if the grill feels like this, and if it sounds like that when they put the steak on the grill, and they are using this specific cut of meat, they should turn it over right about... wait for it... now! This excellent chef may also have warmed the plate and accounted for the time it takes for the steak to reach the table so it is the textbook definition of medium rare when the customer cuts into it and takes the first bit.

How can you use your experience to master a skill?

We can cultivate experiential capital through experiments. I grew about 10,000 extra rhubarb plants one year so I could experiment with different conditions earlier in their growth. I would compost the seedlings once they were too old to be part of the experiment, I simply had no use for that many rhubarb plants.

Through this process, I identified some conditions that improved the results and some that yielded significantly more red plants vs green plants. The red plants are more highly sought after and their stalks sell better, making them more appealing for the commercial grower.

Another way to cultivate experiential capital is through observation. Practice observing. Observe the tiniest details and observe the big-picture patterns. You may have observed that the trees in a certain part of your property tend to grow just a little closer together; why is that? Should I copy that when designing my own plantings for this property?

Observe the insect that never goes on a certain type of plant, you may begin including that plant strategically because of your experiential capital.

I'm a really big fan of experiential capital, I'm always observing, experimenting, and looking for patterns. This form of capital has added as much to my design work as my intellectual capital has. You will find that the same happens as you continue to build your list of experiences intentionally on a daily basis.

Summary

What resources do you have on your property? Do you have a hardwood forest? A softwood forest? The remnants of an old orchard in the back 40? Do you have a pile of scrap wood or metal that can be used for construction projects? Is there a large patch of black raspberries growing along the side of the house that is a problem there, but could be an opportunity if planted elsewhere?

I find that the best way to grasp what capital we possess is to look at our lives from an outside perspective. However you go about it, ensure that you do analyze your capital.

Do you have an inheritance coming? What skills do you have? Who are your neighbours? Which online communities may come together?

Take the time to get to know every form of capital that you have access to well. This can turn a 10-year project into a three-year project while making it more fun and efficient!

Chapter Nine: Threats

Opportunities are great! I embrace opportunities but I have a question for you... What may work against you?

There are many factors that can threaten a project. Remember my renovation with the building inspector issue? The building inspector was a big problem but it wouldn't have been quite as big of a threat if I had more money or had negotiated a longer timeframe with my lender. The combination of the two would have marked an early and expensive end to the project if my lender wasn't understanding.

Now I look to understand my threats before starting a project. When I started this book, I knew that I needed to have the first draft finished in a maximum of two weeks otherwise my schedule would get busy and my attention span would get too short; this would become another incomplete project. I completed the first draft in about 120 hours over eight days. 7 days, 23 hours, and 51 minutes after starting to lay out the outline, I was printing the first copy for proofreading.

There were months of revisions to be made after this first draft. I specialize in applying the contents of this book; it makes sense that I can write it quickly. Editing and revising is where my weaknesses show. I have wanted to drop this project or publish it as-is at least once per week.

I only know this limitation within myself because I have spent time reflecting on my past experiences. I have made a practice of observing and studying myself. I know that two weeks is the longest I will focus on most side projects unless there is some form of accountability or I have a deep reason/motivation to complete the task.

When I completed my financial training I squeezed a full year of training into two weeks because I knew that's how I'll retain the most information and because I knew that I'd struggle if I spread the work over a full year.

What is going to hold you back? Is it an internal factor, something inside you like repeated behaviour, or something external like my building inspector or a cutoff date for applying for a grant?

Threats are OK, the important thing is that we identify them, we recognize them, and we can be honest with ourselves. The risk or threat of having a short attention span and a busy schedule would have been real issues if I had not recognized them. Realizing the threat lead to changing how I wrote this book. I decided that I was going to do something almost entirely unreasonable and write a book in two weeks... Or at least the first draft.

At this point, because I recognized my threats, I was able to turn them into an opportunity. I wrote the first draft of this book in about one week thanks to my threat. Now, on the other hand, I needed to continue to be aware of this threat otherwise there was a good chance that this book would never reach anyone's hands. I needed to press myself to finish writing the book on time and then to keep editing, and formatting long after the two weeks were through.

So what are your threats? What are the things that could hold you back? And how can you turn those into opportunities?

Internal Threats

What are your internal threats? It can be hard to answer these questions. Hard because we may not know the answers, and also because we may not want to face the answers. We don't like to know what makes us weak.

There are many forms of internal threats, we will explore some of them. You may find others that apply to you.

Energy

Energy is one of the first components that comes to mind when I think about managing capital. How do you manage your energy? Are you someone who has vast amounts of energy, can get everything done all at once all the time, or do you struggle to get through each day?

Having an excess of energy can threaten our plans by encouraging us to move too quickly or to take on too much at once. Moving quickly on plans may not seem like a threat, but it can cause issues if we check an item off our list too early.

If you plant your tomatoes too early in the spring the frost may kill them. Likewise, if you put fish into your aquaponic system before the water conditions are right you may kill the fish. If you get excited and flip that breaker to light up the house, you may find that the electrician hadn't finished working on that circuit and that there was a reason for the big red "Do not touch" tag on it.

On the other hand, a lack of energy could be dangerous for reasons that are usually more obvious. You may not have the energy to complete a task. Some tasks can be paused partway through and can be picked up in a day or two or even a year later without causing any harm. Other projects must be completed once they are started otherwise you will be in a worse position than if you had not started at all.

How will you manage the threat that your energy imposes on you? If you have a lot of energy and tend to start projects without taking the time to think about them, you may benefit from having a clearly defined plan including limitations that you impose on yourself intentionally.

If you lack energy, you can hire help, or you can plan stages of work that allow you to take regular breaks without causing issues to the project. With some projects, you can do prep work ahead of time which can turn a big day of work into a much smaller day of work.

For example, if you need to do a lot of baking in one day you can spend the weeks leading up to that day making containers with the dry ingredients for each recipe. On the baking day, you add the wet ingredients and pop them in the oven to bake.

This can also be a great time to invest in social capital by inviting the grandkids over, or maybe inviting the neighbours' kids over, or inviting friends who would enjoy a day in the kitchen with you.

This is not a normal way to go about baking but it is a method that allows you to bake while managing the threat that low energy or chronic pain presents.

Distraction

Distraction is an issue that comes up very frequently with entrepreneurs and homesteaders. I have noticed a pattern that many people who like to start businesses or who want to start a homestead are the same type of people who have a lot of other ideas... always.

If you don't recognize that you tend to get distracted, or that you like starting new projects, it can be devastating to your plans when it happens.

This is the situation that I talked about in writing this book. This is the same situation that I have seen happen with friends who are in business. They get one great idea, develop a plan for it, launch the business, then get excited about the next shiny thing and neglect the first project.

You can turn this into an opportunity in a few ways. One way is the method that I described in this book in which I focus aggressively on the project to complete the project before the next distraction comes along.

Another great option is to surround yourself with people who can work well with your ideas. You can come up with an idea, develop the idea, create all of the systems to make it run perfectly, even participate in the project for the first little while, and then hand it off to somebody else who can continue to run with it. They will keep your vision alive, and they may keep you in the loop so you can follow along and give input as you see the project mature and come to fruition.

The second option is easier when you have a large operation or ageing kids but can be done at a smaller scale. If you have kids, you may help get one of them set up with an egg business, then you might help another one get set up with a beekeeping business, then you might set another one up to propagate perennial plants and sell them.

If you are doing this with your kids, it helps if your kids are involved with the design and with the initial setup. It also can help if they already have an interest in that idea.

One of the methods that I am looking forward to implementing is to allow others to take over projects as subcontractors. Every time I come up with a new idea, I document it thoroughly, including the processes. These processes can be turned into a step-by-step manual that I can hand to somebody as they take over. I would walk alongside the contractor and would include a business plan as well.

If I transition away from beekeeping, I have a list of tasks and their values that I am working on. I may agree to purchase the honey from the beekeeper at a heavily discounted rate in exchange for being the one who provides the bees and hardware.

I can also assign a value to having colony survive the winter. By doing this, I create motivation for the subcontractor to take good care of my bees by giving the contractor a bonus for every colony that survives.

The same thing applies to timber framing projects. I can train someone in the art/trade of timber framing and can either pay them by the hour as an employee or can pay them piece work as a contractor. I ran the numbers for a log cabin that I built in partnership with Trinity Timber Frames (Mallorytown, Ontario). I determined that I could pay someone $100 per log that they milled, cut to length, and installed in the temporary build location.

This method would allow a very slow worker to make a living wage, but a worker who works at the same speed that I did could earn $200,000 per year, one log at a time. I also would earn money from each log without having to do more than check-in and answer questions.

If they stepped back from the project, I could step in and take over again. I could find someone else to take over. I could also let that business hibernate until someone is ready to take it up again.

Whether we are getting a project done quickly, or extra slowly, are hiring help, or are passing a project off to someone else, these strategies turn our threats into opportunities so long as we plan for them correctly.

Health

Sickness and Disease can be considered both internal and external so I'm going to group them all together along with injury.

You may know that you suffer from chronic illness and can factor that into your plan. It is more difficult to know when you will get sick. For example; if you will get a chronic illness after a tick bite or will get diagnosed after going to the doctor because you've been tired for a few months. It is also difficult to know if/when you will get injured.

Health issues aren't as easy to turn into opportunities as some of the other threats that I have mentioned. Sometimes it can be done, other times we need to have more grace for our own needs and limitations.

After breaking my knee, I took six weeks to go to school for a condensed forestry course. After finding out that I was going to need surgery, I took the time to develop my business plan and build my website for Eden's Refuge.

When I came down with COVID-19 for the first time during the middle of a timber framing class I couldn't turn that into an opportunity. All that I could do was sit at home with the curtains closed and a blindfold on as I waited for the migraines to pass.

We can take steps to limit the risk of injury by wearing the proper safety gear, and by paying attention to the task at hand. We can also choose to spend less time doing hobbies that have a high injury rate. We don't want to stop living our lives to limit the threats, so we must be reasonable about balancing the joys of life with the risks of injury.

So, how do we plan for an injury or sickness that we cannot foresee?

I designed my home to be wheelchair accessible. When I knew I was going to have surgery I re-routed the irrigation system so I could turn it on from my back door and keep my plants alive even when I could not walk to the garden.

My long-term plan includes designing my property to require as little maintenance as possible so I can stay there for as long as possible. My business plan includes having an opportunity for other people to pick one basket of food for me, one basket of food for themselves, and one basket of food for the food bank. This will allow me to give away two baskets of food, which of course I will weigh as I am reaching towards my goal of giving away ten tonnes of food per year. This also allows me to make sure my food is harvested even when I can't do it myself.

There are options with insurance. I have seen a various of types of long-term care insurance products and some were very reasonable.

With that product, at the time, I could pay $150/month in insurance premiums until I was 65 at which point the payments stopped while the insurance continued for the rest of my life. If I ever had any injury or illness that affected my ability to complete the tasks of daily living (eating, transferring, bathing, etc), I could collect $750 per week for the rest of my life and that amount would increase with inflation.

One of the perks of that product was that if I was to experience a motorcycle collision this week, I could claim the weekly payment for as long as I needed. Then, when I recovered I could stop the payments but I would still be able to claim against my insurance in the future.

Insurance isn't for everyone; and some insurances are very difficult to qualify for. However, you can self-insure to some extent by creating an emergency fund which will help with short-term health issues. I mentioned a six-month emergency fund in the section about finances which acts as a small insurance policy against emergencies.

Look at the threats that you expect to have, is there something that runs in your family? My Grandfather is in a wheelchair and my understanding is that the cause can be passed down through genetic. I haven't checked if it will affect me, and am choosing not to worry about it. My plans are built so they will not change significantly based on my physical health.

If you are in a particularly sensitive phase of your project during which maintaining your income is essential, you may want to park the motorcycle and hang up your skydiving parachute for those months to limit the possibility of injury.

Our health is valuable, but it can be unpredictable.

External Threats

Natural Threats
Nature is a wonderous thing, but we would be fools to think that it cannot pose a threat to us or to our plans.

We cannot argue against the fact that the climate is changing. We may argue about the causes and about whether it is natural or not, but the climate is changing and that will have an impact on your plans.

Growing up, I remember the one solitary tornado watch that we received. I lost track of how many we had last year, I lost track of how many we had the year before. Tornadoes transitioned from something that happened in movies and in some far-off place called "Tornado Alley" where the childhood version of me pictured constant tornadoes... Adult me doesn't have to drive very far to see the effects of multiple tornadoes.

Consider the changing conditions. Are you in an area that is more likely to get storm surges or tornadoes? Are you in an area that is having droughts that are more frequent and more serious? Are you in an area that didn't have fires historically, but now you fear losing your home each fire season?

Write these threats down, identify them, name them, and plan for them. You can design a property that has fire resilience. You can design systems that will slow the flow of water across your property or that will divert floating obstacles around your home.

We can design a rainwater harvesting system that will meet your needs during those months when your well is dry. We can plant fire-resistant trees to protect your home from wildfires.

Climate change isn't the only threat that nature has in store for us, some threats are constant. This year I lost more than 1000 trees to squirrel and rabbit damage. The solution for this threat is embarrassingly simple and I can promise you that I will not make the mistake of improper fencing again. This does not mean that they won't find a different way to chomp on my plants, but it will not be because I made the same mistake.

I have seen orchards devastated by deer. Do you have dear? If so, where do the deer enter your property from? What do they like to eat? What do they like to avoid? Is there a way to turn these deer into an opportunity?

Can we direct the deer to an area that is filled with weeds that we want to be removed and that they love to eat? Can we design your property with hunting in mind so you can allow the deer in the orchard after you are done harvesting. Allowing the deer to eat the fallen fruit, which will help reduce pest and disease issues in your orchard. Then, as they are finishing their work you can harvest a dear that was fattened on apples, hazelnuts, walnuts, plums, and pears.

Take a moment to think of some other threats that nature can offer us. Soil contamination can be a big one, I wanted to plant a community orchard on some vacant land that my town owns but learned that the only reason why the town has not sold this prime waterfront location is because the soil is contaminated to the extent that it is not financially viable for anyone to clean it adequately to gain approval to build condos.

Currently I'm working on gaining access to those soil test results. Might there be an opportunity for me to plant specific trees and to inoculate the soil with specific fungi that will clean up the contamination? If I can do that, could this land be permanently marked as public land to be used as a community orchard and nature sanctuary?

With that idea, I'm turning a threat against land that I don't even own into a potential opportunity for myself and my community to benefit from.

Pests are natural threats. Some of them aren't naturally part of our local ecosystem but were introduced thanks to globalization and human travel. These are additional threats from nature.

In the permaculture orchard, we like to partner with natural allies like birds to help reduce the threat. There are some other species of insects that our native bird populations will not eat because they are not a part of the local food chain. For these, we may create traps, or come up with other options to confuse or deter the pests.

One opportunity in most situations involving insects as a pest is to support a substantial bird population in the orchard. It is a marvelous experience to walk through an orchard that is full of birds who are all happy to be in what feels like a natural habitat.

What other natural threats are present on your property? Are you in a floodplain? Do you have birds of prey that want to eat your chickens? Do you have birds that fly into your windows too often?

Identify your natural threats and identify their possibilities.

Political Threats

Politics can be a touchy subject. I often avoid conversations about politics because the topic encourages extremes. Rarely in politics do we find the happy medium or the well-thought-out answer that has considered the system as a whole.

Politics are oftentimes this party versus that party. Party A can't recognize the good parts within the plan that Party B came up with to find a solution.

In the homesteading community, there is a lot of fear and anger when it comes to politics. There are a lot of conspiracy theories, and there are a lot of truths. Whatever you believe about politics, they affect you and they can affect your project.

Is the person who is running for the position of mayor supportive of what you're doing, or are they vehemently against anything that has to do with the type of work that you want to do?

One of the projects that I'm working on is introducing community orchards and improving green spaces through increased biodiversity which is shown to increase the positive effects of green spaces on communities, yet my town council keeps voting to sell off green space after green space. Other areas are doing the same as they create additional space for housing.

By following the politics, we can stay ahead of some of these decisions and raise support for the members and causes that are important to us.

You may also find political issues when it comes to something like your mortgage. Interest rates are set by government bodies as a way of controlling inflation. When the economy is very slow and they need to stimulate it, they will decrease interest rates to encourage businesses and consumers to borrow, and spend money which stimulates the economy, adding to inflation to keep it at, or round, their ideal average annual inflation of about 2% per year.

On the other hand, if the economy is experiencing too much inflation, there will be an increase the interest rates to slow down spending and slow down the economy. This can change how much it costs for you to pay your mortgage bill every month. Increasing interest rated can also change how much it costs to finance a vehicle or to pay off your line of credit.

The actions of politicians can also lead to events like wars. I am fortunate to live in Canada where wars don't tend to happen within our borders, but that isn't the case everywhere and that won't always be the case here. I would be a fool to think that Canada will continue for the rest of eternity without experiencing war within its borders. I would be a fool to think that Canada will continue for the rest of eternity, period.

You may live in an area where war is a very real threat. How will this change your plans?

Grants and subsidies are driven by politics. There may be opportunities for grants that are available now but that are threatened by the people who are fighting to get into positions of power. This may mean starting a project earlier than expected so you can qualify for these grants before they end.

The opposite can also be true where the new government may provide grants that will help you finance your projects.

There are many other ways in which politics can affect your property, can affect your goals, and can affect your business. Take some time to think about how you might be vulnerable to political threats. Think about the opportunities as well. Times of low interest are an excellent time to pay down debts quicker because you aren't using as much of your income to pay the monthly interest charges.

Times of low interest also give opportunities to finance projects for those who wish to use debt as a tool. This can allow projects to happen sooner.

Relational Threats

Relationships can be volatile things, they can also be beautiful. I believe that they should be beautiful but the problem with relationships is that people are involved. When people are involved, we tend to bring our imperfections and our insecurities. We tend to bring our incompleteness.

There are many types of relationships in our lives. There are the deep, intimate relationships in marriage. There are the complex relationships of family, whether family is a good word to you or whether it holds a lot of hurt for you.

There are simple relationships too. I recall a sermon on relationships in which the pastor mentioned all of the many ways in which we enter into relationships without even knowing it.

One of his examples was the person who cut you off as you were driving. They may not even know that they cut you off, yet they did. From their perspective, there is no relationship but from your perspective, it's a terrible, horrible relationship with a person who clearly should not have a license... We may be prone to overreacting in this scenario.

The sermon encouraged us to observe how we interact in all of our relationships and to try to act more Christ-like, even in these brief relationships.

The same day I ended up entering into a brief relationship with another driver on the way home from church. It was a snowy day, the type day that is extra slippery as freezing rain was followed by snow.

I was almost home, about one kilometer from home. As I approached a stoplight there were two lanes of traffic in each direction. I noticed that the vehicle in the left lane had stopped early. As I looked at the driver through their rear window I saw them wave another driver in front of them. I laid on my horn and slammed on my brakes before the other driver had even started to make their turn. I knew that the only hope for this relationship to avoid turmoil was for the other vehicle to stay in their lane.

The elderly lady didn't hear me or didn't know what the horn was for. She pulled out in front of me and then hit her brakes as she was right in the middle of my lane, likely a panic response. I moved over as much as I could but still hit the back corner of her car. We both pulled into the gas station and exchanged information.

She was concerned about whether I could afford the repairs or deductibles from the crash because I hit her (so it must have been my fault). The sermon came to mind and I took the time to make sure she was OK. I took the time to inspect both of our vehicles to see what kind of damage was done, it wasn't much, and I was driving an old beater of a car so I wasn't too worried about a cracked bumper.

I learned that she had been in a few accidents recently that were her fault and recognized that another claim on her insurance would likely lead to her losing her insurance or having to switch to a high-risk company which, for an older lady on a pension, is pretty much the same thing.

We called the police; they came and wrote up their report which showed that this lady was in the wrong. At this point, our relationship had the power to change her life. If I demanded to have my car repaired under her insurance (because she didn't have money) she may never drive again. Maybe she shouldn't drive again but seeing it from her perspective I could tell that she couldn't see my car that was coming, she could only see the truck that told her that it was safe to proceed, and she went on faith.

I told her and the cop that I would fix my car myself unless she had to put her repair through insurance in which case, I would like my car repaired. The cop was surprised at this young kid who was letting the older lady get away with the damage to his car. I mentioned the sermon to him in about as many words as I used to describe it in that first paragraph.

Looking back on the situation years later, I can see how impactful that could have been if the roles were switched. If I had trusted the truck driver who said that the way was clear and pulled out and was hit by the little old lady... I was already paying $300.00 a month for insurance just because I was a young male driver. That premium could have doubled or tripled.

How could relationships change your plans? Here are a few examples.

The business partner coming or going. The neighbour with the tractor moving away and being replaced by someone who just wants to rent out the farm and live in the house without a tractor for you to use.

The visitor to your farm who falls, hits their head, and has a life-altering injury then has to sue you even if they are a friend because that is the only way for them to get through the rest of their life.

I enjoy the positive effects of relationships such as when I learn something from someone or when we share a beautiful experience. The best parts of my travels around the world have been the relationships I have found.

We enter into countless relationships every day, and most of these go entirely unnoticed. What are the potential risks? What are the opportunities?

Market Threats
Maybe you decide to start a business. Maybe you're looking to sell a specific crop or to do ecotourism. Maybe that's a petting zoo and some other entertainment, or maybe that is an eco-retreat center.

What happens if your crop becomes the next "big bad food" that everyone is told to avoid? What if you were a peanut farmer just as peanuts were getting banned from schools? What if you were depending on rental income from your eco-retreat center to pay your mortgage. Then a pandemic hits and we go into lockdown for six weeks during your busiest season... and then that six weeks becomes a year without any income... and you still have to pay you mortgage.

What if demand increases? How much are you willing to grow? Is there a cut-off point you will set for yourself off to limit your sales to that quantity, or will you just keep on growing?

Growing is nice, but when we grow too big it can change everything. Are you okay with how things may change? What if someone else recognizes the market that you had a monopoly on, and they start a massive farm right next door to yours to sell their products at prices that you could never compete with?

Diversification is an excellent way to mitigate market threats. Why sell only apples when you can sell apples, and pears and hazelnuts and haskaps and grapes? Not only does this shelter you from market changes but it may shelter you from a season when apples don't do well, or when grapes were all taken out by a fungal disease.

Another way the markets can affect us is when we are purchasing things. The price of a 2x4 tripled a few years ago. Some builders abandoned projects entirely because they could no longer make money, or even break even by finishing a project that they were working on. Some buyers could not pay the new price to finish the job. As a beekeeper, I have noticed the cost of bulk sugar has nearly doubled. These cut into profits.

How could a change in markets change your plans?

Markets are volatile. Fortunately, with good planning, we can adjust, and we can build resiliency into our plan so we have time to shift and adapt as needed.

It is important to recognize what markets we are susceptible to if you want to build a business, or if you want to buy or sell any products.

Summary

You will always experience threats in your life. If you're worried about the air outside you can stay inside but what if your house develops mold or a tree falls on your house?

We will never avoid all of the threats in life, but we can be aware of them and we can have systems in place to deal with the consequences when threats present in our lives.

The goal is to be aware of the threats and to make our plans accordingly. If we are creative enough, we may even be able to turn those threats into opportunities.

It's is not fun to think about all the things that could go wrong and I don't want you to dwell on these things forever. I don't want you to get anxious, or to get discouraged. I want you to see the truth, to accept the realities, and then to make decisions based on what is happening.

If the truth is that you are in too vulnerable of a financial position to be able to invest in your dream homestead, I would rather help you make the most out of your backyard. This will improve your position rather than to design a massive project that may cost you everything and gain you nothing aside from experiential capital which comes from failure.

Look at your opportunities, look at your threats, make an informed decision and know that the worst-case scenario is that you walk away with a little more experiential capital.

Chapter Ten: Strengths & Weaknesses

We all have strengths and weaknesses. Some of us are better at recognizing one or the other. I find that many people struggle to honestly assess their own strengths though.

We either think too highly of ourselves or too lowly. I have done a lot of work in this area, yet I know that I still over and underestimate myself depending on the topic.

Even some of the most self-confident people will overlook some of their valuable strengths. I believe that this is because many of us believe that "If I can do it, it must not be that impressive". I have fallen victim to that mentality many times.

This mindset has led me to believe that my skills are not impressive, but when I take a look at it objectively, I realize that most people can't ride a unicycle down a black diamond trail, or play a dozen instruments. These things feel normal to me because I have been doing them for a long time but as I reflect on my life, I remember that I put a lot of time, and even blood into developing these skills.

Other people have skills that I admire, often they are skills revolving around administrative tasks. I am not great at keeping up with phone calls, emails, scheduling, invoicing, etc, so those skills are ones that I appreciate in others.

There are many strengths in the world, I have written about a few categories below to give you a starting point to start reflecting on them. Take note of what your strengths and weaknesses are. If you have a partner, I invite you to work together on this.

Start by making a list on your own. If your relationship has enough gentleness and trust, share your strengths and weaknesses with your partner and discuss it with them. You can also share your partner's strengths and weaknesses. Do this in kindness, don't use it as an opportunity to touch sensitive spots or old wounds.

If this type of communication is new to you or if you are prone to conflict, consider employing gentle parenting tactics. Saying "I'm logical but you are a dreamer, always starting projects that you don't finish". Consider rephrasing in a way that gets the truth across without poking the bear.

"I'm weak when it comes to dreaming up ideas, but you are good at that and I'm glad you help me try new things. On the other hand, your dreaming is beautiful and you come up with all sorts of business ideas but don't really know how to run a business. I look forward to supporting you through bookkeeping, and building the website as we chase our dreams together" is an example of a very sensitive version of the same statement. You may find a balance between the two extremes.

Your comment may look like "We are both amazing at dreaming up new ideas but we are pretty pathetic when it comes to running a business, we're going to want to get some help with that stuff".

Do it however works within your communication styles, having someone close to you commenting on your strengths and weaknesses can feel vulnerable. This is because it is a vulnerable experience to have someone point out the best and worst parts of you. The benefits are great though as people who are close to us can often see the truths that we miss.

Let's dive into some of the strengths and weaknesses that you may have. These are very loosely categorized.

Physical Skills

Physical Strength	Composting	Foraging
Carpentry Skills	Pruning	Plant Identification
Sewing	Roofing	Fire Starting
Fence Building	Soap Making	Cob Construction
Penmanship	Baking	Seed Starting
Welding	Cooking	Plant Propagation
Electrical	Canning	Seed Saving
Plumbing	Dehydrating	Engine Repair
Glass Cutting	Freeze Drying	Hunting
Concrete	Beekeeping	Trapping
Masonry	Pottery	Breeding Livestock
Smithing	Fermentation	Chainsaw Use
Gardening	Butchering	Tractor Operation

Digital, Business, and Organizational Skills

Web Design	Videography	Packaging Design
Budgeting	Video Production	Data Analysis
Accounting	Conflict Resolution	Compliance
Digital Art	Video Calls	Customer Relations
Marketing	Content Writing	Forecasting
Social Media	Time Management	Market Research
Management	File Management	Negotiation
Legal	Branding	Presentations
Spreadsheets	Logo Design	Public Speaking
Photography	Business Planning	Grant Writing

Experiences

Homesteading	Construction	Hardships
Gardening	Camping	Customer Service
Travel	Performance	Careers

Wisdom & Knowledge

Creativity	Perspective taking	Internships
Storytelling	Courses taken	Certifications
Teaching	Education	Mistakes made

Relational

Communication

Patience

Comprehension

Compassion

Conflict resolution

Collaboration

Listening

Reliability

Persuasiveness

Perspective taking

Expressing emotions

Holding space

Reading body language

Constructive criticism

This is far from a comprehensive list; I have skills and weaknesses that are not on this list. It should, however, be enough to get you thinking.

The Importance of Knowing Your Strengths and Weaknesses

If you want to start an eco-retreat but have no skills in marketing, in hosting, and don't like communicating with people, you may find that you are better off taking a different path. If however, you want to start an eco-retreat and love people and hosting them but are not good at marketing you can trade some of your financial capital (money) for Intellectual and material capital (marketing materials). You may even be able to trade Experiential capital (a free stay at your retreat) to a marketing student for their services.

We can usually find a way to work with both our strengths and weaknesses, but we must understand them first. We must be honest with ourselves and must be honest with our partners.

Chapter Eleven: Creating Your Vision

You've taken the time to identify your values, to understand how sustainable you want your project to be, you have defined your goals and made a plan.

You have learned to assess the eight forms of capital, to look for opportunities, and to recognize threats in your life, both internal and external. You have reflected on your strengths and weaknesses more thoroughly than most people will do in their entire lives.

Now is the time to act... Almost... I promise we're almost there, there aren't many pages left in this book, and some of them are definitions.

You could take the information that you have pulled together to your designer and get started. They are likely to create a great plan for you at this point. There is one more step that some people can benefit from which is to define your vision.

Most designers should be able to help pull coax vision out of you but we'll do that little bit of extra work to define it first to ensure we get the best results.

Picture Perfect

We all have an aesthetic that we prefer. You may want a wild look or a natural but tidy look. You may like straight lines or curvy edges. You may want a modern look, or maybe even an industrial look.

Think about properties that you think are particularly appealing, make a list of them. Find or take pictures of them. These will be useful tools to communicate the general style that you desire.

You can look online for photos as well. Google Photos, landscape design articles or forums, permaculture design articles or forums, homesteading websites, and even Instagram, YouTube, and Pinterest. Be careful with these last three though. I now have them blocked on my phone during productive hours so I don't get sucked down the rabbit hole.

Collect photos, anywhere from one dozen to one hundred. No more than one hundred, and no fewer than a dozen. This range gives a good sample size without sidetracking your project for too long.

Once you have collected photos, scroll through them and see what patterns emerge.

This isn't a big project, just a simple exercise. Once you have completed this exercise intentionally one time, you will have trained your brain to start to look for these design elements that you are fond of.

Here are some things that you can look for and then decide if you actually like those parts of the photos or if they just happened to be in photos that also had something else that you really like.

Wide paths	Statues
Stone edges	Narrow paths
Lots of mulch	Straight lines
Flowers of many colours	Long, curved lines
Flowers of only one colour	Twisty paths
Darker green foliage	Organized chaos
Big leaves	Total chaos
Dainty leaves	Space for reading

Ponds	Tall Plants
Water features	Visual patterns
Waterfalls	Stepping stones
Furniture in the garden	Wood fence
Tunnels	Wire fence
Arches	Rail fence
Gates	Living fence
Mirrors	Pool
Hedges	Hammock space
Stone Mulch	Big Barn
Specific structures	Roof colour
Lighting	Raised beds
Grass	Greenhouse
Steps	Specific trees/plants
Fire area	Specific textures
Dense plantings	Bridges
Low-growing plants	Outdoor cooking space

Again, don't get overwhelmed, that was a long list, but you may only find two or three things that stand out to you. By writing them down you can make sure your design features those elements from the beginning.

This is one of the easiest vision tools for most people to understand and it makes communicating your vision to the designer much easier.

It Just Feels Right

Sometimes what we like most about a property isn't how it looks but is in how it feels. This can be much more difficult to explain. Some designers love to work with feelings.

A space can be designed to inspire a sense of wonder, curiosity, and adventure by incorporating features like a path that turns behind a visual barrier like a tree or hedge. This path disappearing around a corner encourages us to want to explore deeper into the property. Think of the difference between looking down a long, straight road in comparison to looking down that side road that goes over a hill and twists in and out of sight as it follows the valley.

Sideroads like these usually leave me wanting to abandon the current task to explore the mysteries that this sideroad holds.

A space can be designed to give a feeling of escaping into the wilderness. This can be done by using narrow dirt or mulch pathways that weave between trees with a dense understory that invites an abundance of birds to share the space with you.

If you are living in a loud urban area, I like the feeling of turning a corner into a private space with a source of noise like a waterfall. I designed this into my own property (which I then had to remove before renting it out because I couldn't have a pond).

> My yard's design involved walking through a vine trellis tunnel that took me to the back of the private space. At the end of the tunnel was a fruit tree and a fence that was covered with grape vines for privacy, turning right around the edge of the tunnel revealed a "sitting pond" with a stone waterfall. Walking around the mound of flowers and following the stone stepping stones led to the pond entrance.

> Sitting in the pond, I could choose my preferred depth, from cooling off with just my feet in the water to soaking in water up to my neck under the waterfall. All I could hear was the waterfall and I could change the volume with the turn of a control valve.

> Looking around I would see only plants (and my compost bins because I never got around to moving them). To the north was the wildflower hill followed by an apple tree and the grape-covered fence. To the East was a tall cedar hedge and some tropical-looking paw paw trees that framed the waterfall. The South revealed my compost piles which should have been replaced with fruit trees and one of my bee hives. The West showed the trellis which is covered with hardy kiwi vines. In no direction could I see signs of other people.

Your nature space can involve the parts of nature that make you feel the most peace.

Find the words that describe the feeling, or, if you live near an example, take your designer for a visit as you explore it together.

Functions

Is your vision about the way a space functions? Or even about the way that you can function within it?

You may want to create an ecosystem that you can observe and interact with as it heals the landscape.

You may envision how your property can function as an educational garden space, allowing youth and adults alike to learn everything from math to the satisfaction of eating something that they grew for themself. For some people, this may be the first time in their lives that they realize that they can actually do something good.

The Lottery Ticket Exercise

I'm not one for gambling, I agree with financial educator Dave Ramsey when he calls lottery tickets a tax on the poor (ramsey, 2003). Rich people rarely buy lottery tickets. Most of the lottery ticket sales happen in lower-income areas as they are the ones who are desperately hoping for a big win to change their lives.

I have used pretending to win the lottery as a planning exercise. This is best explained through another story.

I was with a friend in his parents' basement, we were playing guitar while thinking and dreaming about the future, about the properties that we wanted to buy, about the lifestyles that we wanted to live. Suddenly he asked if I wanted to split on a lottery ticket so we could dream with the possibility of seeing our dreams become reality.

We are both practical people, yet we can suspend reality to dream of wild situations (like this book selling a million copies so I can start take him to the North Pole and can feed entire communities). We knew that we would not win, we even checked out the odds and looked up how much it would cost to buy enough tickets to guarantee a win (it was an absurd amount of money). We suspended our disbelief and bought one ticket.

We spent the rest of the evening and well into the night sitting on the floor researching and dreaming about the options as $28,000,000 lay on the carpet between us.

It was during this exercise that I defined how big of a property I would need and how much land I want. I picked a specific plot of land, chose a tractor that was appropriate for my needs, and priced out the greenhouses that I wanted to build, and created a list of the plants I wanted to buy. That little piece of paper let us build our dreams without worrying about money.

At the time the jackpot would give us each about $14,000,000. You would think that with two teenagers dreaming their grandest dreams, that money would be gone quickly. We both reached the end of the exercise only having spent about one million dollars, the rest would be invested to provide constant income to do any other projects that we came up with.

Land prices have gone up since then. I believe I was looking at 100 acres for about $400,000. The same dream may cost 1.5 – 2 million dollars today. The plan meant buying everything brand new and all at once.

We realized how attainable our dreams were. A million dollars is a lot of money but we could start with about half of that, and some of that money could be from a property or building loan. At the time I would have needed about $250,000 cash to qualify for the loan for the land and the building. That's a lot of money, but it is something that I could work toward. I could also set my sight a little lower and could have started a scaled-back idea for an amount that I could have saved up in a few years, even as a young guy who was working in a factory.

This exercise is a once per decade activity at most. Spend a few hours having fun coming up with all sorts of ideas and then see what that might look like within the opportunities that you have available to you.

It's a fun tool for dreaming. You can do the same thing without buying the actual ticket, we decided that two dollars each was worth it for five plus hours of entertainment and exploration.

Summary

There are many ways to define your vision. Sometimes the vision is the only information that people have when they approach their designer. I wouldn't make my own design before identifying everything we have mentioned so far in this book so that is my recommendation to you.

The vision project can include the creation of vision boards (a board covered with photos of your inspiration), it can include a poem, it can include a spreadsheet if that is how you prefer to communicate. However you do it, I recommend taking the time to do it.

Chapter Twelve: Communicating With Your Designer

We have gathered a massive amount of information that can be presented to your designer to equip them with much of the information that they cannot gather simply by walking the land and looking at maps.

By doing this work you are equipping your designer with the resources to design your dream project.

There are many ways to communicate with your designer, some designers have preferences, and some know that they communicate best in one format or another. Here are some tools that can help you communicate the initial concept and then to communicate throughout the design process as well as after you have completed the final project.

Before The Design

We communicate in many ways initially. You may meet your designer at a trade show or might call or email them after finding them on the internet. All are fine options, when it comes to communicating what we have discovered through the process of working through this book, we want to make sure we do it well.

This is where some of the pre-work that we have done starts to pay off financially. Saving time in these initial meetings can add up with rates ranging from $150-$500/hour for most designers.

Note: Some designers charge for the initial consultation but will credit you for some of that fee if you choose to proceed with the design.

The reason for this is to reduce the cost of the introductory meeting while preventing people from using it as a cheap alternative to paying the proper rate for a consultation.

Many designers do not provide a credit and that is valid as this is time that they could otherwise be billing for and it doesn't make financial sense to discount skilled labour.

Here are some of the many ways that we communicate.

Face-To Face
Most designers prefer to have a face-to-face conversation, it allows us to have a good conversation, to ask questions, to get a feel for you and your vision, and to walk the land with you.

I prefer to wander the land for a little while before and after the meeting. Walking before the meeting allows me to get to know the land without any vision in mind, a time of observation without intent. The observations that I make can enable me to understand how what we will talk about can fit into the space. The walk afterwards allows me to walk the space to visualize some of the details we discussed, allowing my experiential capital to feel how it would turn out with different overall design patterns.

I was asked to work on a large project recently, it was a commercial-scale operation that would take years to install. On a project of this magnitude, I walked the property once in a very sparse grid to obtain a rough understanding of the land. During the meeting I recognized that they would run into some significant challenges with their initial layout. It would still yield a good product but would cost a lot more money and would take a lot longer to establish.

After the meeting, I spent a few hours creating a detailed map of the property including slopes, waterways, and existing infrastructure. This allowed me to come back to the project manager with an altered plan that would allow them to plant out the first phase in a matter of months in addition to generating rental income in the following year. It included phases of the build that would work within their cash flow.

I also uncovered the motivations of both the landowner, and of the project manager. Their values and visions didn't align and as expected, this caused significant issues as the project progressed. It's currently stalled.

It is important for your designer to have the opportunity to make observations. I believe that it should be a part of any major design work if the land has any complexity to it beyond a steady gradient in a single direction.

When you are together, your designer may have questions. He or she may even point out some of your natural and material resources to see how they may fit into your project.

It is an important time to communicate the things that are most important to you. I would opt for emailing information to your designer before the meeting but if you are sharing it all in person I like to use a different order than the one in which we discovered them in.

This book is arranged so each chapter builds on the ones before it but with the way my brain works, I like to start receiving the information in the following order: values, your range on the sustainability scale, your vision and then your goals. You can add information about the capital that you have available, your opportunities, threats, strengths, and weaknesses throughout your walk or over hot drinks by the fire.

You want a designer who will listen to you and to what you want, not one who wants to put their own dreams on your property.

The in-person visit will usually be followed by email or meeting in person to go over an initial plan.

Video Call

Technology has come a long way, a video call will now allow me to work with people all over the world. You can share your screens with each other to look at pictures and maps. Working exclusively through video calls may mean that the designer never sees the land in person. I'm not a fan of this because we can learn so much from the land.

At the same time, this is a wonderful thing because it means that you can work with a permaculture designer even if there isn't one in your area. It also means that you can choose your ideal designer no matter where they live. This second factor can be important for particularly specialized work where it is more important to have a specific skill set even if it's only for part of the overall design.

With this option, you can email your notes ahead of time which can make the meeting proceed smoothly and much more efficiently. Emailing photos, or providing a link to them can save time that might otherwise be wasted as you navigate the technical challenges of sharing them during the meeting. If you are paying your designer $150-$1000 per hour, time is money, and you don't want to be paying them for technical downtime.

These meetings are often (but not always) followed by an email summary of the meeting. These meeting notes can make sure you are on the same page and provide a record of anything that you agreed to during the call.

Phone Call

This is my least favourite method of having an initial meeting. With no ability to communicate visually, it leaves too much up to chance.

If this is the only meeting before starting a design, I would seriously consider working with someone else. Otherwise, you can send documents and photos ahead of time with clear ways of identifying what you are talking about. A big reference number on each page is great because you can referrer to the correct document easily; "the tree on the right of page 14 was planted by my grandfather and I want to protect it at all costs". Your designer can also ask questions like "On page 9 I see an area on the top-left that appears to have different plants, is this area wet for most of the Spring?"

You need to communicate the same information in the call as in the other methods. I feel as though it is inadequate and should only be used to describe general ideas once the foundational information has been established.

Email is a great tool for sharing information and keeping records. Other designers may disagree, however I would prioritize above phone calls if I could only have one form of communication for an entire design project.

With emails, you can send diagrams back and forth to communicate visually in addition to the text.

During the Design Process

Once you have had your initial meeting with your designer and have given the designer the go-ahead, they will start to work on your design.

Designers vary in their process, the process that I describe here is a common process for most design industries. It's also the same process that I am undertaking right now with the artist for the cover of this book.

Depending on the complexity of your project this may take hours or may take months. The project may be able to be designed from memory and using resources such as satellite imaging and other maps (soil, topography, climactic, etc) or it may require multiple site visits, surveying crews, and even ecological assessments if you are working in sensitive areas or are building on a large scale.

If the designer has all the information that they need, you may not hear from them until the first draft of the design is ready to be presented.

Some designers keep a constant flow of communication whereas others just want to get to work without disruptions.

Initial Concepts

Most designers will make a rough first draft. It's easier to re-design a backyard than a couple hundred-acre project, so bigger projects may be presented to you before undergoing any refinement. This initial design may be as rough as the back of the napkin drawings that we discussed in the chapter "Why Pay for Quality Design?".

The purpose of this first draft is to get the general idea across. If I'm designing a property that has a lot going on I may make a few napkin drawings to show you. With these drawings, we can explore what each might look like in person.

Questions during this conversation may look like this:

Do you prefer this view or that view from your kitchen window?

Would you rather have the outdoor kitchen here, or there?

This is the difference between the two styles that we discussed in the initial meeting, which do you prefer now that you can see both?

Do you want the pond to narrow here to have a smaller bridge or do you want the bridge to be a more dramatic statement?

You mentioned wanting a big nut tree here, these lines show where their shadows will fall during the shortest and longest days of the year. Switching the tree and the storage garden like this would provide summer shade for your sitting area. Does that make more sense to you, or do you want to stick with the original plan? This is what each would look like.

You wanted an accessible design for when you are older. We can make this deck more accessible right away, or we can build it so it's easy to adjust but is much more enjoyable in the meantime.

This is the easiest point in the design process to make changes so if you have any questions or concerns, please ask them so you don't end up with something that you don't like in the final design. You may catch something that your designer missed entirely. You can make comments including:

Is this tree too close to the septic system?

Our utility poles are over here, we are supposed to keep that clear.

I forgot to mention that the farmer has a right-of-way through the property, it's right here.

I don't like this part of the design but I don't know why.

I was thinking about accessibility. I know that I said I wasn't worried about it but as I thought about it more I realized that it is important to me, can we adjust the design to be more accessible?

Take the time to smooth out issues at this point so they don't become ingrained in the design later on in the process. Remember that permaculture design functions in systems so there may be interactions between elements that you didn't consider. These may mean making significant alterations to the plan to change something small that could have been changed earlier with no significant consequences.

This is your project, your designer is the professional and you may have to trust the process, yet you should also trust your gut and remember that it is something that you may live with for the rest of your life.

First Draft of the Design

Once you have worked out some of the kinks with the rough drawings, your designer may complete the final design before contacting you again. They may reach out with small questions along the way, this can be a way to keep you in the loop and maintain a higher quality of social capital in the relationship.

This final draft will be presented to you, it likely will not contain the accompanying information at this point. This shouldn't happen over the phone, at this point you should either be meeting in person or by video call because you are refining the small details.

This is the point when your designer may include some perspective drawings to show details like bridge design, or the view from a specific location on the property.

Speak up about anything that you do and don't like, clear communication is essential.

Final Draft and Delivery

Your designer will then put the polishing touches on your design including creating any accompanying documents. These documents may include:

Plant lists
Plant spacing
Material lists
Detail drawings of elements
Design notes

Timelines
Contact information for contractors
Installation prices (if applicable)
Supplier Information

Delivery can happen in a variety of formats. I like to supply a digital format as well as a high-resolution print. If I know that the clients will want to frame a version of their design, I may provide two copies.

If your designer is doing the installation work, they may or may not include information about material lists or specific species for each plant; they may opt to provide you with a general list of plants, maybe the quantities as well so you know what you are getting into.

After Design Delivery

Depending on your contract, the delivery may be the end of the contract or the business relationship may continue. Either way, it is nice to see a project come to fruition so share updates with your designer!

Designer is Installing

If your designer is installing your project, you won't need to keep them updated because you will be seeing a lot of them. Some designers even offer maintenance services to take care of your property for you. At the time of this writing, I don't have the capacity for that other than to offer some pruning services.

It is the norm not to have installation or maintenance services because they are resource-intensive and can take away from opportunities to do more design work which is where we can do our most important work of helping people transition to a way of life that works with nature.

Installation services can also come with massive overhead costs. Insurance is one cost, but installation work can involve either renting or owning heavy equipment, having labourers (so you aren't paying premium skilled labour rates for basic tasks) and even changing the vehicles that are driven.

I spend most of my time in a fuel-efficient small car that averages 50MPG (Mitsubishi Mirage) but as I transition into installation work I would need to use a truck more often, something that I hope to avoid for as long as possible. This can triple the amount of fuel that I use.

If your designer is doing the installation work they may also hire out portions, for example, if I were doing a large swale or pond project I would look for an excavation company that has the right equipment and the right skills. That two-kilometer-long swale should be accurate in height within an inch over the entire distance. The spillway needs to be even more accurate than that. This isn't something that I would do myself without additional training nor that I would trust most average operators to do.

Your designer may also hire out the construction of buildings and the labour of planting. When you are hiring the designer to do the installation, you are generally hiring them as a project manager, not as the labourer.

Your designer will make sure everything is done according to design and will be able to make last-minute changes according to what they see. For example, if you were having a pond installed that was supposed to be 13 feet deep but at 10 feet there is a layer of clay, your designer may approach you to see if you want to stop at 10 feet because the pond will seal better and may not require a liner or to bring in clay to seal it.

There are benefits to having your designer involved to some level but this isn't the norm.

The designer is not Installing

When the designer is not installing the project, they may include one or more follow-up sessions to answer questions or to help make minor revisions. They may even stop by to check in on how the project is progressing.

In general, unless your contract says otherwise, the delivery of the final design is the end of the business relationship.

I do prefer to have a clause in contracts that allows for some minor revisions down the road, keep this option open if at all possible. Reach out and ask questions if you have them, and ask for clarification if information is needed. No one knows your design better than your designer and no one knows your land better than you do, so work together whenever possible.

If you are installing the project yourself you can choose how quickly to move, you can choose whether to use new materials or to scavenge for secondhand materials and give them another life before they end up in a landfill.

You can choose to hire a machine to dig that big swale or you can head out every morning to dig a little more by hand until you have reached the end.

However you proceed, be sure to read the final chapter titled "After The Design" for tips on how to interact with your landscape over the coming years.

Payment

Payment is the least favourite topic for most small business owners. Most of us have a hard time asking for payments which is part of why so many new businesses fail in addition to many other factors.

Payment can also be a challenging topic from the client side of the conversation. A design can cost a lot of money. Online designers who flat-rate projects from distant lands often start at a few thousand dollars for their design work. Your designer could come in well above or below that, either way, it's a lot of money for most people.

The biggest challenge for clients tends to be knowing when to pay your designer. Here's an outline that is considered good practice in many industries.

Consultation fees for a single meeting are usually paid at the end of the meeting.

Ongoing consultation fees are often paid for on a schedule (weekly, bi-weekly, monthly, etc). This can involve paying for the month ahead, or for the month that has been completed. Paying after every session can be too much of a hassle so a plan is made. Either the client takes the risk and pays ahead or the designer takes the risk and does the work before receiving payment. The longer the period between payments, the greater the risk. Working with someone you trust reduces this risk.

Design work may be paid upon delivery, or, more frequently includes an initial payment of up to 25 percent followed by additional payments. The additional payments may be as simple as one final payment upon delivery or may include an additional payment at a certain milestone. The additional payments are more common during projects that take more than a few weeks to complete.

Installation work is usually paid for in phases with defined milestones and amounts to be paid at that point in the project. An example of this for a timber frame installation could look like:

25% Downpayment
25% When the frame has been cut
25% Upon delivery
25% Upon final inspection

For an orchard installation, it may look like this:

25% Down Payment
25% When plastic mulch and irrigation are installed
25% When trees are installed
25% When bushes, herbs, and birdhouses are installed

Caution:

You should not be expected to pay the entire bill for design or installation work in advance. The practice of billing 100% before starting the work is common among two groups of people.

The first group of people are desperate for money. These people may have good intentions but if they require you to pay the full bill upfront, they may not survive as a business long enough to complete your project or may cut corners to save costs.

The second group involves less honest people who may take your money and run. If you're lucky, they may run before they start work at all but if you are unlucky they may start demolition work and then leave you with a mess.

The rule of thumb is not to pay 100% upfront for anything that is not being exchanged immediately.

Taking an initial payment allows the designer to cover costs. These are bigger for installations than for design work, but both have overhead costs. For example, your designer may pay for access to more detailed maps for your property whereas the installation company may have to buy materials, rent equipment, and pay labour costs.

Payment methods vary between companies. Avoid using credit cards with any small business if possible, you may collect points but the business can pay fees that typically range from 1-5% on all credit card transactions. These fees take a big cut out of their profits. I prefer cash, cheque, or e-transfer because they are usually free or very low cost for everyone involved.

I have also accepted exchanges for small tasks. Someone may offer me some plants that I want and in exchange, I may walk their property with them and share information. This is something that I limit because it can get tricky around tax season "Yes, accountant, for that job I received thirty-seven dollars, some sticks that were cut off of a tree, three chickens, and a roll of old fencing material". I'm tempted to accept that as payment and record it in my books just to see how they react!

Chapter Thirteen: Get Started

Now it's time to start. If you read books the way I do, you may have rushed through the book the first time with the intent of going back and working through the steps... Do it!

The next step will be fun, there is so much to learn and explore. Whether you are designing your property yourself or working with a designer, you will experience hard work, revelations, and hopefully a greater sense of community as you move through the stages of self-discovery, design, and implementation.

Be honest with yourself every step of the way, keep checking in with your values, and don't forget to let yourself dream.

If you are walking through life with a partner, take the time to check in with each other regularly; ensure that you are still on the same page. Don't let unspoken issues turn your dream into a thorn in your relationship.

If you are walking this journey alone, find someone to keep you accountable. This can be a friend, or it can be a member of an online community. Accountability is important in all areas of life.

Get started! When you finish working through the steps in this book you can come back and read the final chapters.

Chapter Fourteen: After The Design

You did it! You have done the work, and you have your design in hand. Maybe you have even gotten the first phase installed. Your journey doesn't end here though. Your journey never ends.

Keep observing what is happening on your property. It will take a minimum of three years for your plants to start to mature and to look the way you intended. Observe what your property has to teach you and don't be afraid to interact with it.

If you see that something isn't working right, observe why, then make a change. You can reach out to your designer for the first few changes but eventually, you will get a feel for it.

If something inside you says that your berry bush should move over six inches, grab your shovel and move it over. Your eye knows what looks good. You are also a part of nature so there is something in us that is designed to interact with it, trust yourself.

Your needs and desires will change over time, you may also come home with a new plant next week that you want to incorporate. Your property is a living, breathing system that will change and evolve with you.

Go, work with nature to make your vision come to fruition.

A Note on Working with the Author

There may be a desire for people to work with me after reading this book. I do enjoy doing design work and I prefer to do it with people who have done the work lined out in this book. This makes me far more likely to want to work with you.

I usually limit design work to properties that I can visit. There is so much that can not be observed through maps and images which limits the quality of design work.

If you have a property near me (at the time of this writing, my business Eden's Refuge, is located in Eastern Ontario, Canada) I will take clients depending on the availability of time. I also visit much of Southern and Central Ontario occasionally.

With my desire to be ecologically sound, I balance the benefits of my work with the effects of travel. A small job of installing a few trees in a backyard is something that I would do locally but it is not worth the ecological damage to travel hundreds of kilometers to do this work unless I can do other projects at the same time.

On the other hand, consulting on a large restoration project or designing an ecologically friendly subdivision would more than warrant flying nearly anywhere in the world.

If you are far from me and cannot find any other permaculture designers I may work remotely with you on a smaller or lower detail project. This may involve more legwork on your end, taking pictures from certain locations, taking measurements, or getting local tradespeople to show up. I may also reach out to other designers in your area if they are suitable for the type of project that you want to do.

Remote consultations are open to people everywhere in the world. It is important to note that while I have visited many areas around the world, I do not have an in-depth understanding (or maybe even a basic understanding) of your local climate, soil types, flora, and fauna. I'm willing to say "I don't know", I am also willing to take a coaching role as you design your property.

Reviews of designs offered by other designers can be done as well. I'm in the process of looking for quality designers who can review my work; there is always room for improvement. It can be a good idea to have someone look over your design, especially if you are using someone who is new to the field or who may be biased in their design work (we all are, but we usually try not to be).

You may reach out through my website www.EdensRefuge.ca or through social media accounts. I am, unlike many businesses, trying to be less available, checking emails a maximum of once per day unless I am actively communicating with someone. I also do not have an assistant so I may take a few days to respond during busy seasons.

Between client meetings, working on projects, teaching classes, and speaking engagements I find that phone calls tend to go to voicemail. I will get back to you but be prepared to leave a message.

If you are working on a privately funded design for community projects may qualify for a discounted rate. I try to make time for some work each month that is priced on a sliding scale for those who otherwise could not afford my services but who would benefit from it.

NOTE: I'm a sucker for really interesting projects, if you have a project that I am interested in and that I can learn something from, I may be willing to travel further. I want to run my business around a seven to eight on the sustainable scale, sometimes that may mean making up for an occasional Zero... maybe....

Glossary

Permaculture A design science following three ethics to create sustainable systems to support life.

Design A plan or drawing created to show the look, functions, or workings of a property, building, or system.

Values Principles or standards that define what a person believes to be important in life.

Ethic A set of moral principles.

Cultivar A plant variety that has been produced through cultivation. Typically reproduced through non-sexual methods examples: grafting, rooting cuttings, etc.

Variety A precisely defined group of plants. Can usually be reproduced from sexual (seed) or non-sexual methods (grafting, rooting cuttings etc.)

Rootstock A plant onto which another is grafted. Can determine characteristics such as cold hardiness, maximum plant size, and disease resistance.

Graft Attaching part of one plant (a scion) to another plant. A non-sexual mode of plant propagation. Common with fruits such as apples and pears to clone a cultivar onto a rootstock with known characteristics.

Swale A ditch created on contour, designed to hold water (not divert it) so it can soak into the landscape. Usually designed with a spillway to prevent failure during heavy rain events.

Spillway An engineered point to allow excess water to leave a system (example: a pond or swale).

Capital Wealth in one of the 8 forms (time, material, social, spiritual, natural, intellectual, experiential).

References

Bacon, F. (1597). *Bacon's Meditationes Sacrae*. Londini: Excusum impensis Humfredi Hooper.

Lawton, G. (Director). (2016). *Greening the Desert Project* [Motion Picture]. Retrieved from https://www.youtube.com/watch?v=xgF9BU4uYMU

Media, P. (Director). (2013). *Miracle Farms, a 5-acre commercial permaculture orchard in Southern Quebec Canada* [Motion Picture]. Retrieved from https://www.youtube.com/watch?v=3riW_yiCN5E

Mgee, C. (2013). The Edge [Recorded by C. Mgee]. On *Permaculture: A Rhymer's Manual*. Fremantle, Western Australia, Australia.

Mollison, B. (n.d.).

Mollison, B. (1988). *Permaculture A Designer's Manual*. Sisters Creek: Tagari Publications.

Ramsey, D. (2003). *Financial Peace Revisited*. Ney york: Viking Penguin.

Ramsey, D. (2003). *The Total Money Makeover*. Nashville: Thomas Nelson.

Roland, E., & Landua, G. (2015). *Regenerative Enterprise*. Morrisville: Lulu press Inc.

Sowards, J. (2021). *Growing Vegetables*. Beverly: Cool Springs Press.

Youtube channels mentioned:

Canadian Permaculture Legacy:
https://youtube.com/c/CanadianPermacultureLegacy

Edible Acres:
https://www.youtube.com/@edibleacres

Stefan Sobkowiak:
https://www.youtube.com/@StefanSobkowiak

Milton Keynes UK
Ingram Content Group UK Ltd
UKHW051256280524
4433835UK00057B/2087